Raise Happy Chickens

Raise Happy Chickens

How to raise healthy chickens and other poultry in your outdoor space

Victoria Roberts BVSc MRCVS

Contents

Meet the author		ix
Only got a minute?		x
Only got five minutes?		xii
Only got ten minutes?		xiv
Introduction		xvi
1	**Chickens**	**1**
	Which breed is best for you?	1
	Buying	9
	Handling chickens	11
	Start-up costs and other considerations	13
	Housing	14
	Routines	29
	Feeding and watering	30
	Health, welfare and behaviour	36
	How to cope with a broody hen	51
	Selling eggs: the regulations	52
	What to do when you want to go on holiday	53
	Breeding your own stock	53
2	**Ducks**	**71**
	Which breed is best for you?	71
	Buying	83
	Handling ducks	85
	Start-up costs and other considerations	86
	Housing	87
	Routines	95
	Feeding and watering	96
	Health, welfare and behaviour	98
	How to cope with a broody duck	105
	Selling eggs: the regulations	105
	What to do when you want to go on holiday	106
	Breeding your own stock	106

3 Geese   117
    Which breed is best for you?   117
    Buying   125
    Handling geese   127
    Start-up costs and other considerations 128
    Housing   129
    Routines   134
    Feeding and watering   135
    Health, welfare and behaviour   136
    How to cope with a broody goose   143
    Selling eggs: the regulations   144
    What to do when you want to go on holiday 144
    Breeding your own stock   145

4 Turkeys   153
    Which breed is best for you?   153
    Buying   159
    Handling turkeys   161
    Start-up costs and other considerations 162
    Housing   163
    Routines   169
    Feeding and watering   170
    Health, welfare and behaviour   172
    How to cope with a broody turkey   177
    Selling eggs: the regulations   178
    What to do when you want to go on holiday 178
    Breeding your own stock   178

5 Other breeds – guinea fowl and quail   182
    Guinea fowl   182
    Quail   184

6 Meat production   188
    General principles   188
    Slaughter   191
    Plucking and hanging   192
    Processing or cleaning   194
    Trussing   197

7 Diseases, problems and general troubleshooting 201
    Free-range poultry diseases   202
    Common problems and some causes   210

	Common diseases by age	211
	Life expectancy	212
	Description of major diseases	212
8	**Cooking with eggs**	**221**
	Favourite recipes	221
	Appendix 1: Vermin control	**229**
	Appendix 2: Transport recommendations	**232**
	Appendix 3: Exhibiting and show preparation	**233**
	Appendix 4: Update on avian influenza and registering of poultry flocks	**236**
	Taking it further	**243**
	Disease prevention	**246**
	Glossary	**248**
	Index	**257**

Disclaimer

Medications and the ability to treat chickens as food producing
animals will differ, depending on the country you live in.
Regulations affecting this should be researched and complied with.

Meet the author

Welcome to *Raise Happy Chickens!*

I was in charge of feeding our Christmas turkeys when I was four years old and my love for all poultry has not diminished in the ensuing years. Wherever I have lived I have kept poultry of some sort, all free-range and mostly pure breeds. At one stage I ran an outdoor pure breed poultry attraction with over 3,000 birds on display. I had an incubator I could walk in to, hatching every week through spring and summer. I can still, after 25 years, tell the breeds of birds from the colour of the chicks, there were that many of them. It was exceedingly hard work, but stood me in good stead for my judging examinations, since until you have kept and bred the pure breeds, judging them can only be theoretical. I have been Secretary of the Dorking Breed Club since 1992, and in 1993 decided that being a veterinary surgeon was a good path to take, albeit qualifying at a rather advanced age, and I have since been able to enhance the welfare of backyard poultry. I edited the *British Poultry Standards*, officiate at major poultry shows and still keep too many chickens. This book is therefore based on my experience.

Only got a minute? Chickens

More and more people are concerned that the eggs they and their families eat do come from happy hens. Eggs are a particularly good source of quality protein.

The battery system of keeping laying hens was developed so that eggs are produced cheaply and in high quantities. Some people would argue that if the hens are laying then they are happy, but these birds have been selected for the highest egg production in the shortest time, so they are programmed to lay in almost any conditions, including indoors in a small cage. If you are concerned about the source of your eggs, have you thought of keeping a few hens in your garden? They do not take up much space or time, and you would know that the eggs produced do indeed come from happy hens. Chickens will also produce manure for the vegetable garden, will weed an area if contained on it and are friendly and entertaining.

You do not need to keep a cockerel in order to obtain eggs and therefore you do not need to worry about noise and upsetting the neighbours. You will need to check on local by-laws, however, before you begin the project.

5 Only got five minutes? Chickens

Once you have decided that keeping chickens is feasible, which breed will suit you? If you sort out your priorities, you will find this out. **You need to decide whether you want lots of eggs per week or only a few; meat; beauty; rare breeds; pets; your vegetable garden weeded.** No single type or breed will fulfil all of these requirements and it is so important that you like the look of the hens you buy as they will become part of the family.

There are many choices of chickens, but the three main types of hens are hybrids, outdoor hybrids and pure breeds. The first type, hybrids, are particular commercial crossbreeds that were originally developed and selected in the late 1940s for the battery cage egg industry, to vastly increase egg production over the pure breeds. They are based on a few of the more productive pure breeds, and tend to be brown, uniform in shape and size, and productive for almost two years, laying 250–300 eggs each year. Examples are Warrens, Isabrown and Hy-line. These are the cheapest as they are reared in large numbers. Beware very cheap ones as these are likely to be at the end of their laying life and rescuing them, although admirable, will lead to heartache as they sicken and die within a few months, but some people feel that, despite this, they are giving them a few months of outdoor freedom.

The original outdoor hybrid, especially developed for free-range systems with hardiness, good feathering and a lifespan of about four years, is the Black Rock. Several others are now available such as the Bovans Nera, Calder Ranger, Speckledy, Columbian Blacktail, White Star and Blue Bell, which are more productive than pure breeds, hardier than the commercial hybrids and provide different colours of birds or eggs. They are a little more expensive than the brown hybrids and best production is 250–275 eggs per year for three to four years.

Pure breeds are the traditional breeds of poultry, developed in a number of countries for various purposes, mostly since Victorian times. They are also used for exhibition and have Standards for shape and colour. On the whole they only lay eggs from March to September. They can produce meat and some of the rarer breeds are particularly beautiful to look at. They live and lay eggs for four to seven years. The light breeds will lay the most but can be a bit flighty and nervous. The heavy breeds lay less and eat more but will produce meat as well. Pure breeds are the most expensive. You could expect 100–250 eggs per year, depending on breed and age.

What about breeds for children? Bantams lay small eggs but eat less and children like them and can handle them. There is a huge choice of bantams in shape, colour and size, some of which are only as big as a pigeon!

10 Only got ten minutes?
Chickens

Good husbandry is really important for the welfare of the chickens you will keep. It is not difficult to make a routine so that the birds have all that they need and are easy to look after. Hens must be housed at night for their own safety. Housing is used for shelter, roosting and egg laying. Some people have just a couple of birds which are in a small house with a run attached and this is moved around a grassed area which provides fresh ground on a regular basis. Others have more birds which may be let out into the garden under supervision or have a permanent run attached. It is very important to maintain vegetation in the run – hens can completely strip a small area in a few weeks, so two runs which can be alternated or a very much larger run gets around this problem. The very worst scenario is mud with no vegetation – battery cages are more comfortable than this!

The hen house must be of sufficient size for the number of birds and have the correct ventilation, near the top of the house to avoid draughts. Perches should be provided as the hens prefer to roost on a perch at night. The nestbox(es) should be in the darkest part of the house and have straw or shavings as litter, never hay due to the moulds present which can affect the heath of the hens.

It is best to purchase a proper poultry drinker since these keep the water clean and enhance the birds' health – dirty water causes disease. A feeder needs to exclude wild birds both from eating the food and contaminating the food with their droppings. It is easy to feed hens nowadays since commercial food with the correct balance of nutrients for laying hens is readily available. Greenery may be added, but it is illegal to feed your hens your kitchen scraps. Whole grain is the other part of the feeding regime which can be scattered on the ground to encourage foraging. Mixed grit should also be provided – the soluble grit helps with shell formation and the insoluble grit is used to grind up the food in the gizzard – remember, hens have no teeth!

Understanding the behaviour of chickens is also important – the word 'henpecked' originates from the pecking order of chickens, which is necessary to maintain social stability in a flock, thus reducing stress. Chickens are naturally cautious, being a prey species, and may take a while to adapt to anything new, so introduce any changes gradually.

If you know how birds work, it will make the instructions on good husbandry more relevant, so this book includes clear diagrams and explanations of the digestive system, the respiratory system and the reproductive system, all of which are different from those in mammals.

Breeding chickens can be another project, but if it is not possible for you to keep a cockerel, fertile eggs may be bought in and either incubated in a small incubator or put under a hen which has gone broody – this is the desire to hatch eggs into chicks and takes 21 days. The broody will then look after the chicks and rear them, teaching them about different foods and how to be safe. If using an incubator, the chicks will need to be artificially reared with a heat source. Either way is successful.

Vermin must be excluded from the hen area. Vermin can be anything that preys on the hens or their food such as mice, rats, crows, magpies, stoats, cats, foxes or birds of prey. Strong and secure fencing is needed and netting over the top of the run is very good prevention against most of the vermin. Foxes mostly hunt at night, so you need to prevent predation by closing the hens in the hut at night – the one night you forget will be the time the fox gets in. If you are going out and will not be back before dark, ask someone else to close the hut for you, or there are gadgets on the market that will close the pophole at dusk.

Keeping chickens is an enthralling hobby and contributes to the quality of the food you feed your family.

Introduction

Keeping chickens is easy, if you follow a few simple guidelines, but, as with all livestock, the responsibility for their welfare is yours. Poultry can take up little time and will benefit a garden with manure as well as being a constant source of fun and entertainment for children and adults alike. In fact, you may well find you are fascinated by the behaviour of these amazing birds – they still fascinate me after over 40 years of keeping them.

Four thousand years is a fair old time for chickens to have been domesticated. They originate from the Red Jungle Fowl (*Gallus gallus*, a small pheasant of Asia) and have provided us with eggs, fresh meat and feathers plus some truly horrible traditional medicines. Domestic ducks are all descended from the lascivious ubiquitous mallard (*Anas platyrhynchos*) and domestic geese from the tame and confiding greylag (*Anser anser*) which, in return for a little corn, would have provided meat, eggs and excellent fletching for arrow flights from the moulted wing feathers when the bow was a common weapon. Turkeys originated in Central and North America and the various pretty colours come from the different subspecies ranging from Mexico up to New England. Guinea fowl are African and, together with the geese, are the 'guard dogs' of the hen yard, shouting loudly at anything strange. It is the Japanese quail that is the domesticated form. They can be kept comfortably in a small area and lay attractively mottled eggs.

Chicken provides 20 per cent of the world's animal protein at a reasonable price – the human race owes a huge debt to the humble domestic fowl. But I want to know where my eggs and poultry meat have come from, don't you? Wouldn't you like the thrill of producing and eating your own free-range fresh eggs? The taste and texture is something everyone should be able to experience. Truly fresh equals less than 24 hours old and a newlaid egg is obvious when you find you have difficulty in cleanly peeling off the shell from a hardboiled egg – the white sticks to it. And what fun to have different coloured or sized eggs for different members of the family.

If you are concerned about how industrial poultry is grown, you can even grow your own poultry meat: you know what it has been fed on, you know it has had a very good life and there is no stress at slaughter as the bird is in familiar surroundings. And the taste is magnificent: what need, then, for those spicy sauces that are promoted to give supermarket chicken some sort of taste?

In this book you will learn which breed or species suits your lifestyle – chickens, ducks, geese, turkeys, quail or guinea fowl – where to obtain them, how to look after them, how to produce delicious eggs for the family and sell any surplus. Eggs are unrivalled and ubiquitous in cooking, so there are some recipes to try, and should you wish to produce your own meat, there is advice on this, plus details on how to prepare a bird for the table.

One small warning – this poultry hobby can be addictive!

1

Chickens

In this chapter you will learn:
- *which breed to choose*
- *how best to look after your chickens*
- *the different behaviours and characteristics*
- *the principles of housing*
- *how to breed hens successfully.*

Which breed is best for you?

Sort out your priorities first: do you want lots of eggs per week or only a few, meat, beauty, rare breeds, pets or your vegetable garden weeded? No single type or breed will fulfil all of these requirements and it is so important that you like the look of the hens you buy as they will become part of the family.

Also, check with your local authority that there are no regulations to prevent your keeping chickens and remember to inform your neighbours, reassuring them that there will not be a cockerel as hens lay without one. Once your new hobby is established, it is amazing how far neighbourly goodwill is enhanced by the gift of fresh eggs.

TECHNICAL TERMS

Autosexing breeds: crossbreeds that have different colours for male and female chicks

Bantam: very small pure breed chicken, very few small eggs per year, children like them, may also be miniature of certain large fowl

Breed Club: collection of enthusiasts for one breed

Broody: hen that sits on the nest all the time to incubate eggs

Chick: up to eight weeks of age

Clutch: a number of eggs laid by one hen until a day is missed (Also the number of eggs a broody will cover for incubation)

Cock or rooster: male over a year old

Cockerel: male chicken up to a year old

Dual purpose: good for both eggs and meat

Free-range hybrid: commercial crossbreed, bred for outdoor conditions

Grower: 8–18 weeks

Hard feather: game birds, used for fighting in 1800s (cockfighting outlawed 1848)

Hen: a laying chicken of any age

Hybrid: commercial crossbreed, bred for indoor conditions

Incubation: keeping eggs at the correct temperature and humidity so they hatch: chickens take 21 days

Pair: male and female

Pecking order: vital social order of poultry

Point-of-lay: 18 weeks (but in any case before laying begins and the best time to acquire chickens)

Poultry Club of Great Britain: responsible for Standards (www.poultryclub.org)

Pullet: a young female chicken before it lays eggs

Pure breed soft feather heavy breed: less good layer, better for meat, may be more docile

Pure breed soft feather light breed: good layer, may be nervous and flighty

Rare breed: pure breed but low in numbers in UK (does not have separate Breed Club)

Standards: published characteristics and plumage colour for each breed

Trio: male and two females

True bantam: very small pure breed chicken, no large fowl counterpart

HYBRIDS VERSUS PURE BREEDS

Hybrids are particular commercial crossbreeds that were originally developed and selected in the late 1940s for the battery cage egg industry, to vastly increase egg production over the pure breeds. They are based on a few of the more productive pure breeds and tend to be brown, uniform in shape and size, and productive for almost two years, laying 250–300 eggs each year. Examples are Warrens, Isabrown and Hy-line. These are the cheapest as they are reared in large numbers. Beware very cheap ones as these are likely to be at the end of their laying life and rescuing them, although admirable, will lead to heartache as they sicken and die within a few months.

The original outdoor hybrid, especially developed for free-range systems with hardiness, good feathering and a lifespan of about four years, is the Black Rock. Several others are now available such as the Bovans Nera, Calder Ranger, Speckledy, Columbian Blacktail, White Star and Blue Bell, which are more productive than pure breeds, hardier than the commercial hybrids and provide different colours of birds or eggs. They are a little more expensive than the brown hybrids and best production is 250–275 eggs per year for three to four years.

Pure breeds are the traditional breeds of poultry, developed in a number of countries for various purposes, mostly since Victorian times. They are also used for exhibition and have Standards for shape and colour. On the whole they only lay eggs from March to September. They can produce meat and some of the rarer breeds are particularly beautiful to look at. They live and lay eggs for four to seven years. The light breeds will lay the most but can be a bit flighty and nervous. The heavy breeds lay less and eat more but will produce meat as well. Pure breeds are the most expensive. You could expect 100–250 eggs per year, depending on breed and age. Bantams lay small eggs but eat less and children like them. Refer to the chart over the page for more information on individual breeds.

A GUIDE TO PURE BREED CHICKEN EXPECTED LAYING CAPABILITIES (LARGE FOWL)

Breed	Egg colour	Numbers per annum	Maturing	Type
Ancona	white	200	quick	light
Andalusian	white	200	quick	light
Araucana	blue/green	150	quick	light
Australorp	tinted	180	medium	heavy*
Barnevelder	light brown	180	medium	heavy
Brahma	tinted	150	slow	heavy*
Campine	white	200	quick	light
Cochin	tinted	100	slow	heavy*
Croad Langshan	brownish	180	medium	heavy*
Dorking	white	190	medium	heavy*
Faverolles	tinted	180	medium	heavy*
Fayoumi	tinted	250	quick	light
Friesian	white	230	quick	light
Frizzle	tinted	175	medium	heavy
Hamburg	white	200	quick	light
Indian Game	tinted	100	medium	heavy
Leghorn	white	240	quick	light
Marans	dark brown	200	medium	heavy*
Minorca	white	200	medium	light
Old English Game	tinted	200	quick	heavy*
OE Pheasant Fowl	white	200	quick	light
Orpington	tinted	180	medium	heavy*
Plymouth Rock	tinted	200	medium	heavy
Poland	white	200	quick	light
Derbyshire Redcap	tinted	200	quick	light
Rhode Island Red	tinted/ brown	260	medium	heavy

Breed	Egg colour	Numbers per annum	Maturing	Type
Scots Dumpy	tinted	180	medium	heavy*
Scots Grey	tinted	200	quick	light
Sicilian Buttercup	white	180	quick	light
Silkie	tinted	150	quick	light*
Spanish (white faced)	white	200	quick	light
Sumatra	tinted	200	quick	light*
Sussex	tinted	260	medium	heavy
Welsummer	dark red-brown	200	medium	heavy
Wyandotte	tinted	200	medium	heavy

* Most likely to go broody. Some colour varieties of breeds lay better than others and different exhibition and utility strains exist, with exhibition strains not being as productive.

It generally comes down to production against beauty. Some people mix hybrids with pure breeds to get a mix of colours of birds and eggs. If planned properly, this is feasible, but there are certain pitfalls to avoid. The pecking order is the chickens' social structure: A can peck B, C and D, B can peck C and D, C can peck D and D is at the bottom. This is how chickens survive peacefully. It needs to be stable in order for production and welfare to be at their best. Once the pecking order has been decided by the flock, upsetting it by adding a bird can create serious problems. Defence of territory is part of this, so it is vital that the whole flock is purchased and settled in at the same time. If it becomes essential to add other birds, let the residents get used to the idea of newcomers by partitioning the hut and run or putting a smaller hut and run within the compound for a couple of weeks so that the newcomers can be seen but not attacked. Removing a bird merely enables those lower down in the pecking order to rise. See page 36 (Health, welfare and behaviour).

Figure 1.1 Light Sussex hen: a very popular pure breed with black neck, wings and tail and white body.

CLASSIFICATION OF BREEDS

Soft feather: heavy
Australorp
Barnevelder
Brahma
Cochin
Croad Langshan
Dorking
Faverolles
Frizzle
German Langshan
Marans
New Hampshire Red
Orpington
Plymouth Rock
Rhode Island Red
Sussex
Wyandotte

Soft feather: light
Ancona
Araucana
Rumpless Araucana
Hamburgh
Leghorn
Minorca
Poland
Redcap
Scots Dumpy
Scots Grey
Silkie
Welsummer

Hard feather
Indian Game
Modern Game
Old English Game Bantam
Old English Game Carlisle
Old English Game Oxford

True bantam
Belgian
Dutch
Japanese
Pekin
Rosecomb
Sebright
Serama

Asian hard feather
Asil
Ko Shamo
Kulang
Malay
Nankin Shamo
Satsumador
Shamo
Taiwan
Thai Game
Tuzo
Yakaido
Yamato-Gunkei (True bantam)

Rare soft feather: heavy
Autosexing Breeds: Rhodebar,
  Wybar
Crèvecoeur
Dominique
Houdan
Ixworth
Jersey Giant
La Fleche

Modern Langshan
Norfolk Grey
North Holland Blue
Orloff
Transylvanian
  Naked Neck

Rare soft feather: light
Andalusian
Appenzeller
Augsberger
Autosexing Breeds: Legbar,
  Cream Legbar, Welbar
Braekel
Breda
Campine
Fayoumi
Friesian
Italiener
Kraienkoppe
Lakenvelder
Marsh Daisy
Old English Pheasant Fowl
Sicilian Buttercup
Spanish
Sulmtaler
Sultan
Sumatra
Vorwerk
Yokohama

Rare true bantam
Booted
Nankin
Ohiki

Rare hard feather
Belgian Game
Rumpless

Commercial hybrids
- ▶ *Lots of eggs*
- ▶ *Hens all look the same*
- ▶ *Need to be replaced after two years*
- ▶ *Cheapest*
- ▶ *Not very hardy*

Free-range hybrids
- ▶ *Good numbers of eggs*
- ▶ *Come in different colours*
- ▶ *Live a year or two longer than commercial hybrids*
- ▶ *More expensive*
- ▶ *Hardier*

Pure breeds
- ▶ *Some eggs, most of the year*
- ▶ *Different colours and patterns*
- ▶ *Live four to seven years*
- ▶ *Most expensive*
- ▶ *Hardiest*

Top tips
- ▶ Check local regulations.
- ▶ Choose your breeds carefully according to the end product required.
- ▶ Do not add new birds to an established flock without careful thought and planning.
- ▶ When your birds arrive, shut them in the hut with food and water overnight then quietly open the pophole in the morning – if they find their own way out without pressure they will find their own way back again that evening.

Buying

Study the health points and the diagram and follow the biosecurity guidelines below so that you know what to look for when going to buy birds. Reject any that do not come up to scratch: don't buy them because you feel sorry for them – they will be nothing but trouble.

BIOSECURITY FOR FREE-RANGE CHICKENS

▶ *Isolate new stock for two to three weeks.*
▶ *After exposure at an exhibition isolate birds for seven days.*
▶ *Change clothes and wash boots before and after visiting other breeders.*
▶ *Change clothes and wash boots before and after attending a sale.*
▶ *Keep fresh disinfectant at the entrance to poultry areas for dipping footwear.*
▶ *Disinfect crates before and after use, especially if lent to others. However, it is preferable not to share equipment.*
▶ *Disinfect vehicles that have been on poultry premises but avoid taking vehicles onto other premises.*
▶ *Wash hands before and after handling chickens.*
▶ *Comply with any import/export regulations/guidelines.*

These are common-sense measures that can easily be incorporated into a daily routine.

POSITIVE SIGNS OF HEALTH IN CHICKENS

▶ *Dry nostrils.*
▶ *A red comb (some breeds have naturally dark ones).*
▶ *Bright eyes (colour varies with breed).*
▶ *Shiny feathers (all present).*
▶ *Good weight and musculature for age.*
▶ *Clean vent feathers with no smell.*
▶ *Smooth shanks.*
▶ *Straight toes.*
▶ *The bird is alert and active.*

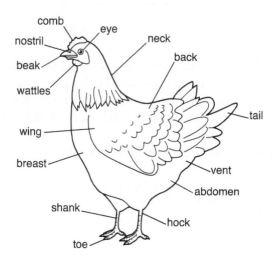

Figure 1.2 The external anatomy of a chicken.

AGE OF ACQUISITION

Point-of-lay (POL) is about 18 weeks of age with the fast-maturing hybrids. POL with pure breeds could be 26 weeks or more. It is better to obtain the hens before they begin laying so that the laying cycle is not upset by a change of environment. If you transport a bird that is in lay, it is very likely to stop laying for a few weeks. You should expect eggs about two to four weeks later.

Pullets may be obtained at an earlier age but will need different feed in order to grow properly (a grower ration, see page 63).

WHERE TO BUY

Check all stock using the health signs above before purchase.

- ▶ *From advertisements in smallholding magazines such as* Country Smallholding, Smallholder, Fancy Fowl *and* Practical Poultry.
- ▶ *From private breeders who exhibit at poultry shows.*
- ▶ *From private breeders. Ask to see the parent stock.*
- ▶ *At small sales. Talk to the breeder.*

▶ *Large sales as prices can escalate and there may not be any history with the birds such as age, health status and how they have been reared.*

▶ *Adverts in local newspapers may be genuine or dodgy, for example they may be selling end-of-lay battery hens as POL free-range birds – if it sounds good, go to see the birds before purchase.*

▶ *Car boot sales and the internet.*

Handling chickens

Insight

In order to maintain her place in the pecking order, a hen will disguise the fact that she is not feeling well.

Handling on a regular basis is very important as it is the only way to tell if a bird has lost or gained weight. Even when really thin their feathers disguise this fact, so handling will give a vital early clue to any problems. Loss of weight and excess weight can be assessed by feeling the pin bones either side of the vent: they are sharp if the bird has little fat and well padded if too fat. The distance between them will indicate if the hen is laying: three vertical finger widths between the bones indicates production and less than two indicates the reverse.

BEST HANDLING METHOD

If you first come across a hen in a cardboard box, slide your outstretched hand, palm up and fingers spread, blindly into the top of the box, then under her from the front, with her breast resting on your outstretched palm and her legs between your closed first/second and third/fourth fingers. Your other hand should be placed over her back to balance her as you lift her out of the box. Take the weight on your forearm and hold her close to your body with her head pointing towards your armpit, leaving your other hand free to inspect her. Be firm but don't squeeze the body tightly as this may temporarily harm the breathing mechanism (see pages 42–43 for internal anatomy). This principle of holding applies to all species and all sizes of poultry – the bird is balanced and comfortable and the mucky end is away from you. The hip of any poultry will dislocate with horrifying ease if a bird is held by one leg. Do not hold them by the legs upside down.

If your first introduction to a hen is not in the confines of a box or crate, begin the handling procedure by practising in the dark with a very dim torch when the hens should be on their perch. If you move quietly and slowly, talking to them all the time, you will not startle them and you can then pick one hen off the perch with both hands around her wings and body, facing towards you. Then continue by sliding one hand under her as above.

You are likely to need to catch a hen during daylight hours, so obtain a fishing landing net as this can (with practice and the aid of a fence or wall) be dropped over the hen and you can then pick her up as above. This is much less stressful for both you and the hen than chasing her around the pen or garden, as she will be able to run and jink much faster than you.

Hens can become really tame if handled on a regular basis – some like sitting on a warm lap and watching television!

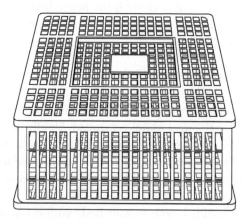

Figure 1.3 A poultry crate.

Start-up costs and other considerations

The choice is huge for dedicated hen houses, from the superbly designed and long-lasting down to the minimal that will last one or two years. Of course, the price reflects this. There is also the converted garden shed or existing stable, see below.

START-UP COSTS

Henhouse and run (e.g. for 6 hens)	£250	($375)
6 hens @ £6	£36	($54)
Drinker and feeder	£30	($45)
Plastic dustbin for feed storage	£5	($7.50)
Total	£321	($481.50)

VARIABLE COSTS PER ANNUM

Feed: allow 100 g of layers' pellets per bird per day;

6 hens will eat 25 kg (one bag) in 6 weeks (9 × £6)	£54	($81)
Wheat 100 kg (@ £4 per 25 kg)	£16	($24)
Grit, shavings or straw	£20	($30)
Greens in winter	£10	($15)
Total	£100	($150)

It is easy to waste feed, so get the proper feeding equipment, whether galvanized or plastic.

INCOME/BENEFITS

▶ *Maximum production from a hybrid is 300 eggs per year, about five eggs per week per hen, total 30 eggs per week. If the family eats 15 eggs per week, the remainder can be sold for £1–2 ($1.50–3) per dozen, giving £65–130 ($98–195) income, which will go towards covering the cost of the feed.*

- *Manure is a valuable commodity: about eight 25 kg bags of manure will be produced from six hens in a year and can either be used in your own garden or sold for about £2.50 ($3.75) per bag.*
- *Weeding of a vegetable garden or allotment.*
- *Hours of observation and enjoyment!*

Housing

TECHNICAL TERMS

Apex roof: two slopes
Droppings board: removable board to catch droppings placed under perches
Fold unit: movable self-contained house and run, may or may not have wheels
Free-range: access to grass in daylight
House, hut, coop, cabin: henhouse
Litter: dry and friable substrate on the floor
Nestbox: to lay eggs in
Pent roof: one slope
Perch: roosting place
Pophole: low exit door
Run or pen: fenced exercise area, usually grassed
Shavings: livestock woodshavings for litter, also to line nestbox
Skids: used to move larger huts
Straw: usually wheat straw as barley straw is too soft
Ventilation: must be at roof level and above heads of birds
Window: replace any glass with wire mesh

Housing and space will depend to a great extent on how many birds you decide to acquire. A small garden may only support two hens, whereas a larger area could have a dozen. The basics of housing are the same no matter how large the house, and commercially produced housing is usually well designed and built. Manufacturers advertise in the country magazines such as *Country Smallholding, Smallholder, Fancy Fowl* and *Practical Poultry*.

If building your own, buying secondhand or converting a garden shed, consider the following points:

- ▶ *Is there sufficient ventilation with no draughts? (Ventilation should be just under the roof.)*
- ▶ *Are the nestboxes easy to get at from the outside of the house for egg collection?*
- ▶ *Are they sited in the darkest part of the house (under the window) as the hen will want to lay in the most secret place she can find?*
- ▶ *Are the perches wide enough (5 cm or 2"), long enough (at least 23 cm or 9" per bird), and the right height (see below) so the hens use those and not the nestbox to perch in?*
- ▶ *Is there a droppings board to make cleaning out easy and to keep the floor clean?*
- ▶ *Depending on the system in use, is the house easy to move?*
- ▶ *Is it fox-proof and other vermin-proof?*
- ▶ *Has the timber been treated for longer life with a non-toxic substance?*
- ▶ *Is there felt on the roof which may harbour red mite? If so, replace with clear corrugated plastic.*
- ▶ *Has the house been designed by people who keep chickens and therefore understand their requirements?*
- ▶ *Is the house substantially built so your investment will last for years?*

Poultry housing is used by the birds for roosting, laying and shelter. The welfare of the birds is entirely in your hands and certain principles must therefore be observed.

SPACE

Floor area should be a minimum of 30 × 30cm (12" × 12") per bird (large fowl) or 20 × 20cm (8" × 8") for bantams. If you can give them more space then so much the better, bearing in mind they will be spending time in the henhouse sheltering from the rain and wind. Perches should allow a minimum of 23 cm (9″) for large fowl and 15 cm (6″) for bantams, but also remember that they help to keep each other warm in harsh weather.

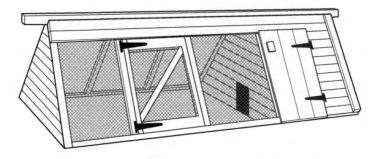

Figure 1.4 An ark for a pair of bantams.

VENTILATION

Correct ventilation is vital to prevent the build-up of bacteria and condensation. It should be located near the roof to ensure there are no draughts. It is often more difficult keeping the house cool than warm. If hens get too hot they become aggressive and can begin feather pecking each other, potentially leading to cannibalism.

Insight
In summer when it gets light early and the hens want to be out, hang up some stinging nettles in the house to keep them occupied until you get up.

Window
A mesh window would normally be located near the roof with a sliding cover to allow for adjusting the ventilation. Glass can break and does not help the ventilation. One window is best as the house can then be sited with its back to the wind. The amount of light affects egg laying with 14 hours being the optimum. If you are adding artificial lighting in winter, set it to come on with a timer in the early morning. This will allow the birds natural evening

twilight to choose their roost. Alternatively, fit a 15-minute dimmer switch to be used in the evenings to achieve the same ends.

NESTBOXES

Nestboxes are best located in the lowest, darkest part of the house as hens like to lay their eggs in secret places. Size for large fowl is up to 30 x 30cm (12" x 12") or 20 x 20cm (8" x 8") for bantams with one nestbox per four hens. Communal nestboxes with no partitions are useful as sometimes all the hens choose just one nestbox and queue up or all pile in together, which is when eggs get broken. Make sure there is outside access for you to collect hen eggs. Litter in the nestboxes can be shavings or straw (not hay due to moulds). If the hens are sleeping in the nestbox, check the perch is higher than the nestbox and, if it is, gently move the birds to the perch in the dark or block off the nestboxes until they begin laying.

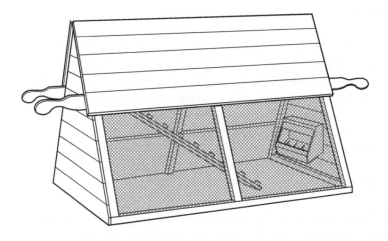

Figure 1.5 A two-storey ark with feeder and drinker attached.

If they are laying on the floor of the hut, pin vertical 5 cm (2") strips of black binbag (bead curtain effect) to the front of the nestbox to make it darker and more inviting. Pottery eggs can be placed in the nestbox for encouragement.

PERCHES

Even for bantams, perches should be broad – 5 cm (2″) square with the top edges rounded. They should be the correct height for the breed so birds can get on them easily and have room to stand up on them (see above for spacing, but allow 30 cm (12″) between perches if there is more than one). Make sure they are higher than the nestbox otherwise the hens will roost in the nestbox instead. Positioning a droppings board under the perches that can be removed easily for cleaning will keep the floor of the house cleaner as hens do two-thirds of their droppings at night. You can also health check the droppings for colour and consistency more easily. Normal droppings are brown with a white tip. Every couple of hours a hen will void the contents of her caeca (blind gut which ferments herbage) which is lighter brown and a bit frothy.

POPHOLE

This is a low door so that the hens can go in and out of the house at will. The most practical design has a vertical sliding cover which is closed at night to prevent fox damage. The horizontal sliding popholes quickly get bunged up with muck and dirt and are difficult to close. There are light-sensitive or timed gadgets available which will close the vertical pophole for you if you have to be out at dusk.

SECURITY

The house must provide protection from vermin such as foxes, rats and mice. 2.5 cm (1″) mesh over ventilation areas will keep out all but the smallest vermin. You may also need to be able to padlock the house against two-legged foxes. If your garden is exposed and you live in a windy area, make sure the house is low and long (rather than tall and thin) with a low centre of gravity, or tie it down with ropes in bad weather. If the house is rolled by the wind, both the birds and the structure will be damaged, possibly irrevocably.

MATERIALS

Timber should be used for the frame, which can then be clad with tongue and groove, shiplap or good quality plyboard. If the timber is pressure treated by tanalizing or protimizing it will last without rotting. The roof must be sloping to allow rain to run off. This could be either apex (two slopes) or pent (one slope). Avoid using felt if possible as this is where the dreaded red mite parasite breeds. Onduline is a corrugated bitumen that is light and warm, therefore reducing condensation. Plywood can be used if it is treated with Cuprinol or Timbercare, which are the least toxic wood treatments where birds are concerned. (If you must use creosote, even the modern varieties, leave the house empty for at least three weeks to allow the toxic fumes to disperse.) To protect the plywood roof further, use corrugated clear plastic instead of felt as it lets the light through and deters the red mite which prefers dark places. Square mesh is best used on the window and ventilation areas as it is fox-proof.

Sectional construction is best for ease of moving. If building your own bear in mind that the house may need moving – a common mistake is to make it so heavy that several people are required to lift it. Wheels are available from many manufacturers to make moving their houses easy enough.

Recycled plastic has recently come on the market as a poultry housing material. It is rot-proof, light, easy to clean and less likely to harbour parasites. The Eglu is one recent innovation as a start-up henhouse in plastic.

LITTER

Woodshavings for livestock are the cleanest and best, straw is cheaper but check that it is fresh and clean, not mouldy or contaminated by vermin or cats. Do not use hay because it harbours harmful mould spores that will give the hens breathing problems. Litter is used on the floor, in the nestboxes and on the droppings board. Deep litter means having a friable and deep substrate such as shavings and/or straw and the hens scratch this over on a daily basis. It helps to break down the faeces and if properly managed has little smell. It is changed annually and makes exceedingly good compost, with the high nitrogen content of the manure breaking down the tough shavings before it is added to the vegetable garden. It should never be used fresh as it will scorch plants.

FLOOR

The floor can be solid, slatted or mesh. Slats should be 3.2 cm (1¼") wide with a 2.5 cm (1") gap between. If slats or mesh are used, make sure the house is not off the ground otherwise it will be too draughty. Slats or mesh make for easier cleaning. If the house has a solid floor, raise it off the ground about 20 cm (8") to deter rats and make a dry dustbathing area. If a deep litter system is opted for, such as using a stable or shed with an earth floor, spread plastic under the shavings or litter to prevent damp and previous organisms from rising.

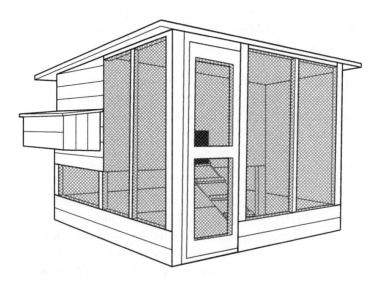

Figure 1.6 A two-storey hut and run, using all ground area.

CLEANING

Weekly cleaning is best, replacing litter in all areas. The best disinfectant which is not toxic to the birds is Virkon. This is a DEFRA (UK Department for Environment, Food and Rural Affairs) approved disinfectant that destroys all the bacteria, viruses and fungi harmful to poultry. If using the deep litter system, this should be cleaned out once a year, but the nestboxes should be cleaned weekly and the perches checked for a build-up of faeces.

PEN OR RUN?

Some houses come with a detachable run or you can make your own, preferably with a net or solid plastic roof to deter wild birds from defecating near your poultry. If movable, then the grass will stay in good condition and will be fertilized. If static, be aware how very quickly the run can become either a sea of mud or bare of vegetation, so make it larger than recommended or divide it into two so that each side can be rested on a regular basis. If the area is on the small side, consider placing strong, 2.5 cm (1″) square netting over the grass so that the hens can still eat it but do not destroy the roots. They will appreciate a separate dustbath if this method is used, as the netting will prevent this natural activity.

Some people put down wood chips (not bark as this harbours harmful mould) to maintain free drainage. A large run should have netting over the top to prevent wild bird access. If this is done, the feeder and drinker can be put in the pen or run, otherwise they should be inside the house to discourage wild bird access – not only does DEFRA consider wild birds a risk to domestic poultry due to disease, but they will steal a huge amount of the chicken food and magpies will quickly learn to take eggs, even from inside the henhouse.

If you are gardening, the hens will love to help you find worms and insects, but they are best let out under supervision as they have a tendency to try to replant everything. The very small bantams do least damage. Free-range in a domestic situation usually means daylight access to grass, not necessarily total freedom.

Beware poisonous plants such as laburnum, laurel and nightshade, but if you have children you won't have these in your garden anyway. Daffodil bulbs are toxic, so be careful of these, although most poisonous plants taste horrible to hens. Unless the covered run is a large area, don't attempt to plant shrubs inside it as the hens will soon dig these up. Clematis, honeysuckle, berberis, pyracantha or firs can be grown on the outside of the run both for shelter and to enhance the area.

If you want to weed an allotment, use a fold unit which is a house and run combined that can be moved to a fresh piece of ground as soon as the hens have done their job, possibly daily, which means any droppings can be incorporated immediately as there will only be a few. If the hens are contained within the fold unit (with the feeder and drinker hanging in the run part) they will efficiently weed and manure an area of your choice and leave your precious vegetables alone, as well as being protected from the fox.

Figure 1.7 Barred Wyandotte hen: more curvy than the Sussex and not such a good layer, but the feather pattern is similar to the Maran, which lays dark brown eggs.

BUY OR MAKE?

If housing is bought from a reputable manufacturer and meets all the basic principles then that may be the quickest and easiest method of housing your birds. If you wish to make housing yourself, keep to the basic principles and remember not to make it too heavy as you will want to move it either regularly or at some stage. Remember to make the access as easy as possible for you to get in to clean, catch birds or collect eggs. Very occasionally secondhand housing becomes available. If you choose this option beware of disease, rotten timbers and the inability to transport in sections.

Should you already have a suitable garden shed and wish to use this, make some A-shaped, free-standing, 5 cm (2″) wide perches, about 30 cm (11″) high and long enough that all the birds can roost in comfort. Make enough nestboxes to suit the size of birds – you will have to go into the shed to collect the eggs – and place these under the window (darkest place). Ideally, make a wire inner door so that the main door can be left open in warm weather without wild birds getting in. Hang the feeder and drinker off the floor, about the height of the smallest bird's back. Attach a board to the base of the doorway to avoid litter being scratched out, ideally make this removable for easy cleaning. Check that there is some ventilation near the roof. If not, drill some 5–7 cm (2″–3″) holes and cover them with small wire netting or remove a section of boarding and cover this with similar netting.

TYPES OF HOUSING

There are two basic types of housing, movable and static. Movable pens are good as the birds get fresh ground regularly. Some have wheels which makes moving them easy for anyone and some have the house raised off the ground so that hens have more grass area but are still contained. Triangular arks were developed to prevent sheep jumping on housing in the days when different stock was kept together, but the shape of the ark can damage the comb of a large cockerel or hen. A disadvantage of movable pens or fold units is the limit on the size and therefore the number of birds kept in each one.

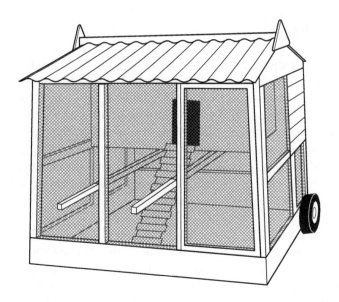

Figure 1.8 A two-storey hut and run with wheels for easy moving.

Static or free-range housing needs to be moved occasionally in order to keep the ground clean around the house – alternatively use slats or flagstones in the highest traffic area. Using this method the hens are allowed to roam freely or are contained within a fenced-off area but you must protect their feeder and drinker from wild birds. Larger huts may have skids so they can be pulled by a vehicle to move them. Remember that tall, thin houses are unstable in windy areas, so go for something low and broad based. If a sliding or hinged roof is incorporated there is no need to have the house high enough to stand

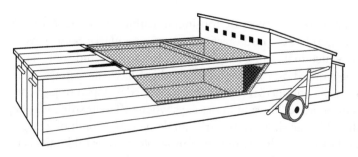

Figure 1.9 A fold unit with wheels.

up in. When a free-range house has a solid floor and is raised off the ground to discourage rats and other vermin from hiding under the house, make sure the nestboxes are attractive as the hens are liable to lay under the house if their nestboxes are not adequate. When moving pens on a daily basis it is useful to have feeders and drinkers attached to the unit so you don't have to take the equipment out and put it all back again each time you move it.

If you have a stone or brick building that you want to use for hens this is obviously not movable so you may wish to go for a deep litter house, even if only in the winter. Figure 1.16 shows where to put the relevant equipment and you can let birds out or contain them in harsh weather. Either have hard standing outside the building that can be kept clean, or place slats round the entrance to keep the feet of the birds cleaner and prevent disease-inducing patches of mud. If the building has a half door, keep the top half closed to deter wild birds, however a wire mesh inner door with a pophole in it will be more effective.

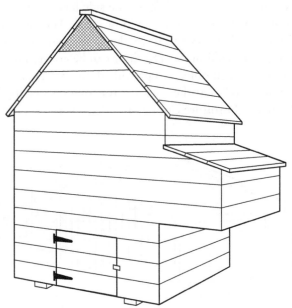

Figure 1.10 A free-range house with good ventilation and outside nestboxes.

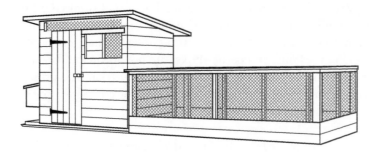

Figure 1.11 A house on skids with a detachable run.

Alternatively, fix the door so that it can be secured partially open so the hens can just get in and out.

The aviary system of keeping chickens is another method, but requires a larger garden. The aviary can be one free-standing house or a line of them with rotational access to grass. The principle of an aviary is that it is spacious, at least one side has mesh, the roof is solid and the walls can be either solid or mesh. The birds are protected from the elements and have plenty of fresh air without contamination from wild birds. The floor can be covered in straw, sand, gravel or wood chippings to provide drainage. There should be a shelter with perches at the back for roosting plus laying boxes, feeder and drinker. Furniture can be provided for entertainment

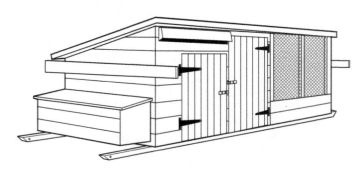

Figure 1.12 A house with a fixed run on skids.

such as branches to climb on, but don't be tempted to plant inside the aviary unless wire protection is given to the plants as hens are destructive. Climbers can be planted on the outside to enhance the area and shrubs protected with wire can be in the grassed area.

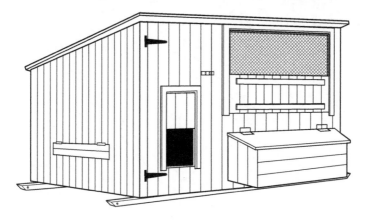

Figure 1.13 A house for up to 50 birds with external access, droppings board, skids and outside nestboxes.

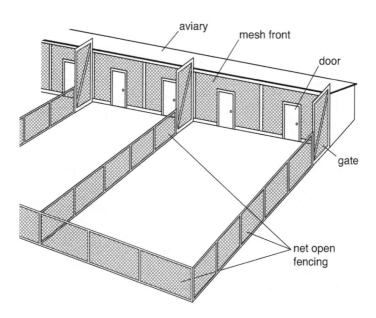

Figure 1.14 A line of aviaries with rotational access to grass.

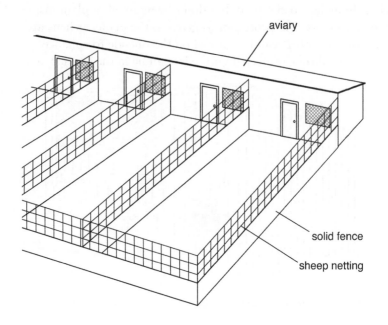

Figure 1.15 Aviaries with individual runs. note the solid fencing divisions with sheep netting above to prevent territorial disputes.

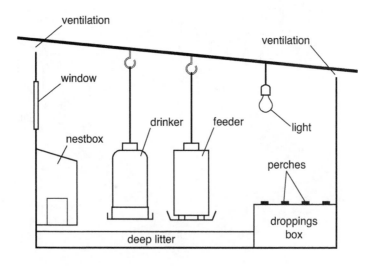

Figure 1.16 Example of a deep litter arrangement.

Routines

Give the hens as much time as you can as you will enjoy your hobby more, but take just a few minutes daily to check them.

- ▶ *Let the birds out at dawn in winter and at about 7 am in summer.*
- ▶ *Change the water in the drinker.*
- ▶ *Put food in the trough or check the feeder has enough food for the day.*
- ▶ *Collect any eggs.*
- ▶ *Add fresh litter if necessary.*
- ▶ *Scatter a little whole wheat either in the house or in the run.*
- ▶ *Observe the birds for changes in behaviour that may indicate disease.*
- ▶ *Shut the pophole before dusk, checking all the birds are in the house.*

WEEKLY ROUTINE

This is the time to get to know your birds and keep a closer check on their health.

- ▶ *Clean the nestbox and/or droppings board or floor. Put cardboard under the shavings to make the procedure easier; it will also compost down.*
- ▶ *Scrape the perches.*
- ▶ *Put the dirty litter in a covered compost bin.*
- ▶ *Wash and disinfect the drinker and feeder.*
- ▶ *Check that mixed grit is available.*
- ▶ *Handle every bird to check their weight and condition.*
- ▶ *Deal with any muddy patches in the run.*

Feeding and watering

TECHNICAL TERMS

Ad lib feeding: hens able to feed at any time (protect this from wild birds)

Automatic drinker: connected to header tank or mains with valve

Automatic Parkland feeder: feed is accessed by pecking a coloured bar

Breeder pellets: to give to the adults four to six weeks before they lay the eggs for hatching

Chick crumbs: high protein small-sized feed for chicks

Feed bin: vermin- and weather-proof bin such as a dustbin to keep feed in

Feeder: container for food to keep it clean and dry

Free-standing drinker: container for water to keep it clean

Gizzard: where food is ground up using the insoluble grit (see also page 42 for internal anatomy)

Grower pellets: for growing chickens

Layer pellets: for laying hens

Mash: a commercial feed not really used for free-range as wasteful

Mixed corn: wheat and maize combined, only useful in cold weather as very heating

Mixed grit: needed for the function of the gizzard

Pellets: a commercial ration or feed in pelleted form, grower or layer composition

Quill drinker: triangular shape, fed from header tank with nipple drinkers along base edge

Scraps: household food – this should only be given to hens if it is raw vegetable matter or a little stale brown bread

Spiral feeder: metal spiral at base of large bucket, pellets accessed when spiral pecked

Wheat: fed whole as a treat

DIGESTION

Insight

'As rare as hens' teeth' is a common saying, which is true, as hens' food is ground up in the gizzard.

Food and plant material is delicately picked up with the beak, which is made of keratin, rather like human nails. As food, the texture is more important than the taste as they have few taste buds. Hens use their feet to scratch the soil and expose insects for food. Colour vision and shadows help the hen to find food.

When hens feed, the crop is used to store food. This is then passed to the proventriculus (acid-producing stomach) and then to the gizzard where it is ground up. The food is passed back and forth several times between the gizzard and proventriculus and then, when fine enough, passes through into the small intestine. Nutrients are absorbed from the small and large intestine and the liver then filters nutrients and toxins from the food, carried in the blood. The caeca (paired, blind-ended portions of the gut) is where a certain amount of fermentation of plant material takes place and the droppings from this area are voided about one time in ten, and are slightly frothy and paler in colour than normal droppings which are brown with a white tip. The white tip is the urates (bird form of urine), which in birds is normally solid. Figure 1.26 (page 42), the internal anatomy diagram, shows the placing of the kidneys which are long, paired, and held firmly within the protection of the spinal column.

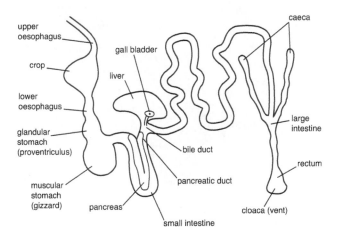

Figure 1.17 The digestive system of a hen.

EQUIPMENT

The correct equipment for commercial feed will reduce waste and keep contents clean. There are many designs of drinker but the main principle is the vacuum: plastic drinkers are often designed so that the upside-down top is filled with clean water and the base is then slotted onto it, the whole drinker then inverted – the vacuum prevents the water from pouring out – with a rim for the hens to drink from. More durable, but more expensive, galvanized drinkers work slightly differently in that the base is filled with clean water and the top pushed over it and slotted in, with the water trickling out of a small hole. Some people like to add cider vinegar to the water to slightly acidify it and make the bacteria (added every time a chicken drinks) less likely to multiply, but don't use galvanized drinkers for this as the vinegar can react with the metal. If a stressful incident is anticipated, add soluble probiotics (beneficial bacteria) to the water or feed a little plain yoghurt (no fruit, no sugar) in the food to help protect the intestine from the stress hormone which tends to strip out beneficial bacteria, leaving spaces for harmful bacteria to lodge.

Figure 1.18 Rhode Island Red hen: this is a bantam, but the square shape of the body denotes a good layer.

A feeder with vertical bars is best to prevent the hens from hooking food out. Also, choose one with either a sloping lid or a bar that revolves to prevent them from perching on it. If birds have access to grass they will not need extra greens but if there is not enough grass in winter, hang up some cabbage stalks, nettles or brussels sprouts plants in their hut. A swede cut in half and impaled on a blunted nail previously driven into a block of wood will provide much entertainment for the hens.

FEED AND WATER

It is important that only balanced feeds from reputable sources are used. Cheap feed will be of poorer quality. Feeding scraps tends to upset the balanced ration which has been proven over many years, but green vegetable matter is appreciated in the winter and to be able to call the hens over with the reward of a small piece of fruit or stale brown bread will be very useful. The compound rations can be fed either as pellets or meal/mash. The meal can be fed dry (this may be wasteful and sticks to the beak making any water quickly foul) or as a wet mash. When mixed as a wet mash it should have enough water added so that when pressed in the hand and released it should just crumble away. Pellets and dry meal can be fed in *ad lib* feed hoppers but wet mash must always be fed freshly mixed as it goes rancid very quickly.

Insight

Clean water and mixed grit should be available at all times.

Empty drinkers in hot weather are as bad for the hens as frozen water in winter – they dehydrate quickly. Flint (or insoluble) grit is needed to assist the gizzard in grinding up the food, especially hard grain. From four weeks before laying commences, oyster shell or limestone grit should be provided *ad lib* to help with the formation of egg shells. Light breeds start to lay at about five months and heavier breeds at about six months. Large fowl will eat about 110–170g (4–6oz) per day of the ration and bantams need around 50–85g (2–3oz), according to size. Wheat (and a little maize in cold weather only) can be offered as a scratch feed to keep the birds active. If they are not free-range, green feed is always welcomed by the birds, but hang up vegetables and nettles to get the most benefit from them otherwise they just get trampled.

Store feed in a vermin-proof and weather-proof bin to keep it fresh. Check the date on the bag label at purchase as freshly made feed will only last three months before the vitamin content degrades to an unacceptable level.

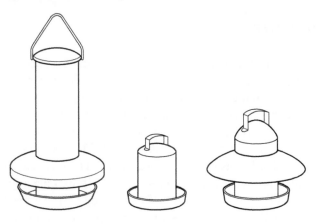

Figure 1.19 Eltex equipment (l–r): tube feeder, drinker, hat feeder.

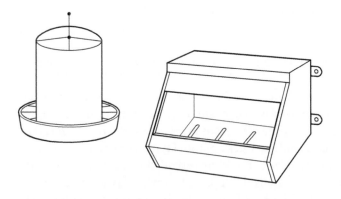

Figure 1.20 Different types of feeder: gearwheel and wall. Prevention of feed wastage: bars to stop hens flicking food out.

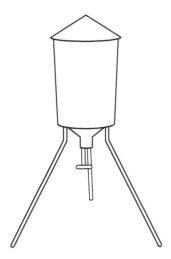

Figure 1.21 A parkland automatic feeder.

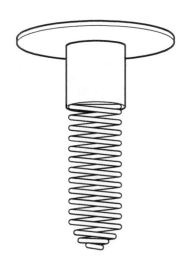

Figure 1.22 A spiral feeder.

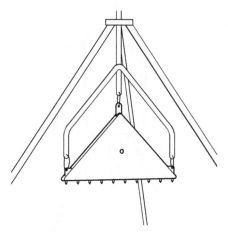

Figure 1.23 A quill drinker.

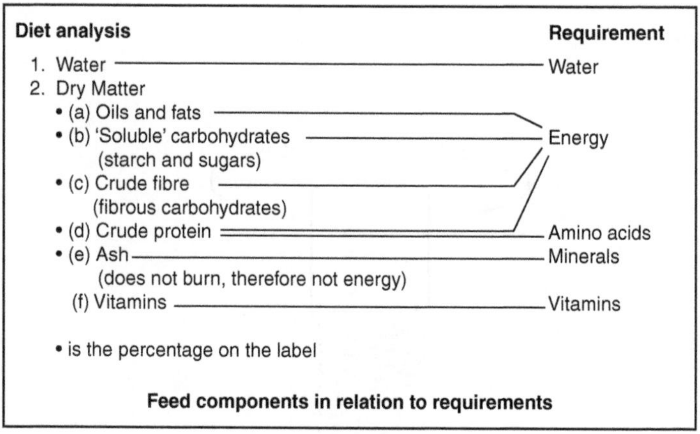

Diet analysis   **Requirement**

1. Water ————————————————— Water
2. Dry Matter
   • (a) Oils and fats ————————
   • (b) 'Soluble' carbohydrates ————————— Energy
       (starch and sugars)
   • (c) Crude fibre ————————
       (fibrous carbohydrates)
   • (d) Crude protein ══════════════════ Amino acids
   • (e) Ash————————————————— Minerals
       (does not burn, therefore not energy)
   (f) Vitamins ————————————— Vitamins

   • is the percentage on the label

Feed components in relation to requirements

Figure 1.24 Feedbag label contents.

Health, welfare and behaviour

The welfare of the birds is entirely in your hands, so if you
follow the guidelines in this book you will have healthy and
happy birds – they will repay your care by giving much enjoyment
and fresh and delicious eggs.

Many common poultry conditions or diseases can be avoided
if something is understood about the poultry's behaviour.

Poultry are creatures of habit – life is safer that way – so any
change in routine can upset them. This can range from a sudden
snowfall – when they will not venture outside as the ground
has changed colour – to a sudden change from meal to pellets,
one of which does not look like food. This is not stupidity: their
confidence is easily dented which, of course, is part of the survival
mechanism. The key word here is 'sudden': they will cope with
most changes if they are gradually implemented.

Poultry do not have much sense of taste or smell but they
are sensitive to texture. They use shadows to spot potential
food items on the ground and anything falling or moving is
immediately investigated. All birds have colour vision and hens
are particularly attracted to red, hence the red bases to chick
drinkers. Unfortunately this also means that any fresh blood is
also attractive, which can lead to cannibalism, no matter how
large the free-range, but once the blood has dried the danger is usually
past. Hens make bad gardeners – they seem to be convinced that
everything you have planted is upside down. What they don't
scratch up, they dust bathe in to help remove external parasites.

The word 'henpecked' is so much part of our language that most
people use it without thought to its origin, but the pecking order,
as we have seen, is a vital component in maintaining the stability
of the flock in hens. With or without a cockerel, hens have a strict
hierarchy and territory which changes only if a bird is removed or
is sick. How then do you return a recovered bird to a small flock?
With great care and supervision, as their memory is very short and
the whole structure has to be reshuffled. The recovered bird must
be fully fit in order to reclaim her place. If hens fight, the cockerel
will usually break it up. Should you wish to add say, three or four
new laying hens to an established flock, the least traumatic way to
do it is to turn their liking for routine against them and put old and
new into a fresh henhouse so no one has established a territory.

Young, unconfident laying hens cannot be expected to start or continue to lay if being bullied by the rest of the established and confident flock: the youngsters can quite easily be killed by the matriarchs. Old English Game are best left to the experts as even chicks out of the same clutch will kill each other.

Unlike waterfowl, hens can be immensely stupid in rain. The older ones will quickly decide if it is only a shower or a lasting downpour and run for home, but young birds are liable to stand around in the wet looking miserable. They will become cold, especially those with thin skulls and crests such as Polands and those with woolly feathering such as Silkies.

Anything overhead or flying is a potential predator – one bird will often alert the breeder to a sparrowhawk or buzzard and then there is a general warning buzz. Hens do not see well in the dark, so they feel safe in a dark enclosed space at night and put themselves to bed before twilight. They do not herd well, even in daylight, but two long bamboo canes can work wonders in getting them all to go in the desired direction.

The birds you have bought will probably have been vaccinated against some of the more common poultry diseases (but ask in any case) and as long as they are fed correctly, stress is kept to a minimum and they are wormed about twice a year, they should be healthy. Never buy a hen that has a runny nose or noisy breathing. This is caused by an organism called mycoplasma and can be carried by wild birds. It can be controlled but not cured by antibiotics. In Chapter 7 there is a diseases chart. It is a summary of common poultry diseases found in small flocks. It is essential to involve your veterinary surgeon (take this book with you) if you have problems with your poultry and although some wormers and louse powder can be obtained through licensed outlets (such as agricultural merchants), most drugs and medicines are only obtainable through a vet. Wash hands after handling medicines and observe the withdrawal instructions on the labels of drugs, so do not eat eggs or birds when medicines are being given.

If medicines are given in water, make no other water available. Most diseases are management related, for instance rats and mice carry some diseases as well as all those carried by wild birds, therefore many diseases can be prevented by good management.

If you acquire birds that have not been vaccinated, should you vaccinate them? If there is a problem in your area with a specific disease, then it would be sensible to vaccinate them against that disease, but if you only have a few birds with no problems in the area, it will probably be safe not to vaccinate. The vaccines come in industrial sizes and so 95 per cent of them would be wasted in order for your few birds to be protected. If there is a local disease problem, then it is money well spent.

Plastic string is very dangerous. If you see string hanging from a hen's beak, do not under any circumstances pull, but cut it off and hopefully the bird will eventually pass it through. String (such as that closing feed bags) can get wrapped around the leg or foot of a bird causing swelling and ultimately gangrene, so get into the habit of picking it up.

If you understand how a bird works, you are more able to keep the husbandry standards high as the reasons for so doing make for healthier birds.

Insight

A healthy respiratory system in poultry depends very much on good ventilation so that disease can be prevented: damp, stale air will quickly cause problems.

▶ *High levels of ammonia from the litter stop the removal of phlegm and so invite bacteria and viruses to multiply. If you can smell ammonia in the hen house, there is too much, so clean out the litter more often and increase the ventilation.*
▶ *Hens are omnivorous and enjoy catching and eating mice, but the disease risk is high from rodents. Not having teeth, pieces*

of food are ground up in the gizzard, so avoid old long grass as this can impact and kill, as can polystyrene (e.g. ceiling tiles) which they just adore pecking at, or pieces of plastic string.

Figure 1.25 Partridge Wyandotte hen: another bantam but shows the intricate pattern of this colour.

▶ Muddy areas encourage harmful parasites to breed so put down slats or move the hut more regularly.
▶ Hens have evolved to scratch around in the dirt, but over a wide area. Problems occur if they are kept on the same small area of ground all the time, which then becomes 'sour' and harbours harmful parasites and other pathogens.
▶ Feathers are good insulators and it is sometimes harder to keep birds cool in summer than warm in winter. Birds that are too hot will hold their wings out from their body and pant.

TECHNICAL TERMS

Abdomen: belly

Airsac: air storage areas which then act as bellows

Airspace: very small when newlaid egg

Albumen: white of egg

Caeca: paired blind-ended parts of intestine where some plant fermentation takes place

Candle: in a dark room, shine a small torch on the broad end of the egg in order to see inside

Chalazae: two strands of thick white which help to keep the yolk the right way up (especially if fertile)

Comb: fleshy protruberance on top of head, varies in shape and size with breed

Crop: food storage area near top of breast

Dustbath: box with dry soil or ashes

Dustbathing: a hen rolls around and flicks dry soil over her feathers to help remove parasites

Germinal disc: where the yolk is fertilized, needs to stay on top, hence the chalazae

Gizzard: strong muscular organ containing small stones or grit to grind up food

Hock: 'elbow' of the leg

Infundibulum: funnel-shaped structure where the ovum is deposited, chalazae added

Isthmus: shell membrane added here

Lungs: do not expand but have air pushed through them

Magnum: albumen (white) added here

Moult: annual replacement of feathers

Oviduct: tube where the egg is constructed

Ovum (plural ova): one egg yolk in the left ovary (right ovary not functional)

Post-ovulatory follicle: tissue remaining after the ovum has passed into the oviduct (ovulation)

Preen gland: small gland just above tail producing oily substance

Proventriculus: acid-producing stomach

Shank: leg between hock and foot

Sinus: area just below eye
Spleen: small, purple, shiny and round
*Sunbathing: a hen lies with outstretched wing and leg towards
    the sun*
Syrinx: voicebox of birds at lung end of trachea
Trachea: windpipe
Uterus/shell gland: shell added here
Vagina: egg passes through here in one minute
Vent: anus
Vice: detrimental action
Wattles: fleshy protruberances under beak

Figure 1.26 indicates where the internal organs are placed. This will
help you as a common mistake of novices is to feel the hard gizzard
and think the hen is egg-bound (an egg stuck in the oviduct).

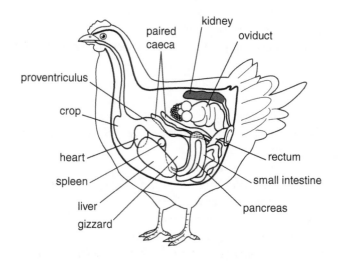

Figure 1.26 The internal anatomy.

Figure 1.27 gives some idea of the very different respiratory system
in a bird – the lungs do not expand as in mammals, but the air
is pushed through them in one direction by the movement of the
ribs and airsacs in a bellows motion. This is one reason why it is
dangerous to squeeze a bird – it can be prevented from breathing.

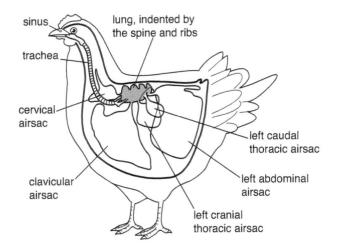

Figure 1.27 The respiratory system.

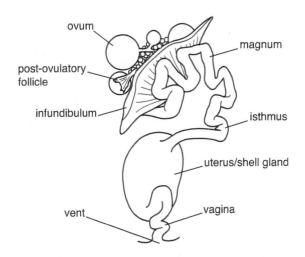

Figure 1.28 The reproductive system.

This is also why songbirds do not appear to stop for a breath when singing, as air is continuously moved through the syrinx (voicebox) which is at the lung end of the trachea.

Figure 1.28 shows the area where an egg is made. It takes different times for the egg to pass through the different areas of the oviduct, with the addition of the shell taking the longest time:

▶ *15 minutes in the infundibulum (fertilized here if cockerel available plus chalazae added)*
▶ *three hours in the magnum to add albumen (white of egg)*
▶ *one-and-a-half hours in the isthmus to add shell membrane*
▶ *20 hours in the uterus/shell gland for shell deposition plus pigment*
▶ *one minute in the vagina, which is extruded out past the vent to avoid the faeces.*

Insight

In total it takes 25 hours to lay an egg. This explains why hens do not lay every day, as the hen will ovulate 30 minutes after laying. If it is dark when this happens, she then misses producing an egg the next day.

The hybrid hens produce eggs after a slightly shorter time, hence they lay on consecutive days for longer. Brown eggs have pigment placed on the outside of the shell which can be scratched or washed off. The only egg that has pigment all the way through the shell is from the Araucana and it is blue/green. The shell is porous and so if eggs are to be washed, water warmer than the eggs should be used (with Virkon, a disinfectant) so that the shell membrane expands and blocks the pores. If water colder than the eggs is used, the shell membrane will shrink and draw in any bacteria on the shell.

If fertile eggs are needed, the semen storage glands in the oviduct ensure the cockerel does not need to be with the hens every day. This is important if the hens are exhibited, so that their feathers do not get damaged. If another breed cockerel has been used and a different cockerel is wanting to be used (for instance, to change to a pure breed) you will need to wait two weeks for the eggs to be true to the new cockerel, due to the semen storage capacity. It is not possible to tell from the outside if an egg is fertile, only after a week when it has been incubated (see candling, page 59), therefore cockerels should not be run with hens if eggs are to be sold in case the eggs are not stored properly and a fertile egg begins to develop an embryo.

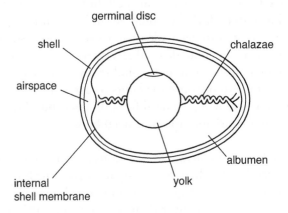

Figure 1.29 An egg.

Figure 1.29 shows the composition of an egg. This is important as it is how a fresh egg is determined. The airspace is very small in a newlaid egg and gets progressively larger as the egg loses moisture through the porous shell (see page 58 for incubation). When a fresh egg is cracked onto a plate, the thin white and the thick white are easily distinguished and the yolk sits in a defined dome on top of the white. The chalazae can be seen. A not-fresh egg will appear to have only one type of white, no chalazae and will be flat when viewed from the side. Sometimes there may be a small brown mark in the white – this is a tiny amount of blood (known as a meat spot) and is not harmful, it just doesn't look nice. Commercial eggs are candled to remove any with meat spots. Fertile eggs are candled to check on the development of the embryo. The colour of the yolk depends on feeding and commercial feed has additives to enhance yolk colour. Carotenoids in green plants are the basis of yolk colour, so outdoor birds usually have darker yolks in the summer and paler in the winter.

In a hut 2 m x 2 m (6' x 6') a 40-watt bulb lit to create 14 hours of light in total including daylight would be sufficient. It is important that hens have twilight to persuade them to go to roost, so fitting a 15-minute dimmer is a good idea, otherwise when the lights go out they are stranded on the floor and would really prefer to perch. It suits some people to organize the lights to come on in the morning instead, removing the need for a dimmer, but this may create noise and disturbance for your neighbours.

It is one of life's great unanswered questions as to how male birds keep their semen fertile when their testes are inside the body and birds' body temperature is higher than a mammal (mammal testes have to be a degree or two lower than body temperature, and hence are external). Evolution, of course, has helped, and having internal parts ensures a certain amount of protection.

A change in behaviour may signal the onset of a problem or disease, so time spent observing poultry is never wasted. They are fascinating creatures in any case, providing much entertainment with their antics and characteristics. They will become as tame as you want them to be and are very quick to learn that a spade means worms. Bantams are particularly good if children are involved as they are easier to handle, the eggs are instantly recognizable and the hens can be bred from for extra interest, with the cockerels being much smaller and somewhat quieter.

The skin of hens is thinner than that of mammals and is normally white (e.g. Sussex), yellow (e.g. Rhode Island Red) or black (e.g. Silkie). Leg colour depends on plumage colour and breeding cockerels have a red blush down their legs. Yellow leg colour can fade if a hen has been laying well as the pigment goes into the yolk.

Hens are programmed to moult all their feathers once a year. It takes three to four weeks for a new set of feathers to grow. The hybrids sometimes moult only every 18 months and stress can cause a premature moult. The old feathers are pushed out by the new ones coming through and in a specific pattern, usually not all at once, although the better hybrid layers may look like a hedgehog for a week or two. The new feathers have blood in them until they are fully through and then the blood is resorbed. Other hens like to pick at these blood-filled quills, so if you can spot a culprit doing this, isolate her until the other's feathers have fully grown.

If feathers are missing off the back of a hen, it is usually due to the feet of the cockerel when he is mating. If her feathers are missing on the front of her neck or under her tail, it will be her friends who are pulling them out. Remember the alpha hen is not allowed to be pecked by the lesser ones so she may even be the perpetrator!

Hens spend much time preening in order to maintain the properties of feathers such as insulation and a certain amount of waterproofing. They will dustbathe in order to try to get rid of feather parasites and for this they need a dry area that is easily available in the summer but in winter, a box with dry soil or ashes in is much appreciated. Sunshine is important for the maintenance of vitamins, so hens will sunbathe with one leg and wing outstretched – seemingly dead until something disturbs them.

Should you need to contain a bird behind a lowish fence, it is permissable to clip the feathers on one wing, making flight unbalanced and therefore difficult. To do this, cut the feathers as in Figure 1.30. Hens are very good at jumping, however, and may still get over a fence, the grass always being greener on the other side.

If you have done some breeding, you will be left with some cockerels that no one wants. A good end for these is in the pot, see page 189. At least you will know how they have lived and what they have been fed on.

If, however, you have a stock cockerel that has become a bit territorial – perhaps he rushes towards you when you enter the run – at the first sign of this pick him up and cuddle him. This will thoroughly confuse his aggressive instinct and defuse the situation. If you retaliate with a broom or dustbin lid, he will merely be encouraged as you are playing his game.

An aggressive bantam cockerel is amusing, but a large breed can be positively dangerous with small children. Once the habit has set in, I do not know of a way to cure it apart from one involving a roasting tin.

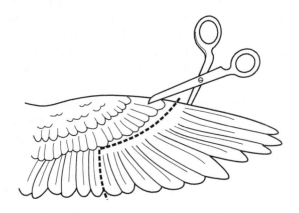

Figure 1.30 Wing clipping.

PARASITES

Internal parasites

These are common in hens that roam outside during the day. They are always on the lookout for insects and worms and some of these can contain harmful parasites. It is easy to control these by giving a worming powder called Flubenvet to the hens in their feed. If the birds are in an ark and are moved regularly, they should be wormed two to three times a year. If they are on the same ground all the time, this interval should be shorter as they may be re-infecting themselves frequently. Flubenvet has nil withdrawal time for eggs, so it can be used at any time. It is obtainable from

48

your vet, also see page 202. Some of the internal parasites cannot be seen with the naked eye, so worm your birds on a regular basis.

External parasites
There are several types of external parasites, all of which should be dealt with as dustbathing will only remove a few:

▶ **Lice:** *These live on the bird, are yellow, about 3 mm (0.12")
long and lay eggs (nits) at the base of feathers, usually under
the tail, so this is a place to look on a regular basis. Control is
with louse powder based on pyrethrum. Not life-threatening,
but reduces production.*
▶ **Mites:** *The red mite is 1 mm (0.04") long, nocturnal and sucks
the blood of hens at night, while living in the hut during the day.
Hens become anaemic with the comb and face pale and can die.
Red mite can be controlled by spraying the hut with various
licensed products when the birds are outside, but the nooks
and crannies, especially under a felt roof, are difficult to get at.
Careful application of a blowtorch is just as effective, although
time-consuming. Either treatment will probably need several
applications. The red mite is grey when it is hungry and will take
a meal off a human if it gets the chance. They can live for up to
a year without feeding so beware secondhand henhouses! If the
hens refuse suddenly to go into the hut at night, suspect red mite.*
▶ *The northern fowl mite is a relative of the red mite but lives
and breeds on the bird all the time. On a white bird, this is
easy to see as a dirty mark on the base of the feathers or under
the tail. Around the vent is the most common place to find
these so, again, check here regularly. The birds become anaemic
within a few days of being infested due to anaemia and can die.
The life cycle in warm weather of both types of mites can be as
short as ten days, so vigilance is really important. Treatment is
pyrethrum-based louse powder or an avermectin (ask your vet)
as nothing is licensed for northern fowl mite.*
▶ *Scaly leg mite burrows under the scales of the legs making
raised encrustations and is very irritating for the bird.
Treatment is by dunking the affected legs in surgical spirit
once a week for three weeks.*

VICES

Sadly, hens can acquire vices but these are often due to management and husbandry problems.

▶ *Egg eating happens if an egg gets broken in the nestbox. This could happen if there is not enough litter, the nest is too small or the shell of the egg is weakened for some reason or even missing. Hens know that eggs are extremely good food and will take any opportunity to consume an egg once a shell is broken. Prevention is best, such as covering the front of the nestbox with vertical binbag strips, providing enough litter and enough grit, but if the hens (and it might only be one, look for egg on her face!) are egg eating there are several things you can do. Do not under any circumstances follow the old wives' tale and waste a mustard or curry-filled egg in the hope that it will put them off as hens do not have the taste receptors for hot spices. Put several ping-pong balls, golf balls or*

Figure 1.31 Silver Dutch cockerel: the smallest of the bantams, the Dutch do well in quite small spaces

pot eggs in the nestbox and on the floor of the hut and these should put them off. Find out why the eggs have been breaking.

▶ *Feather pecking can begin if the growing birds get too hot or if there are too many in a brooder. Reduce the temperature. You may also have to take the worst offenders to another brooder. In adult birds, feather pecking is often caused by boredom so hang up some nettles to give entertainment. If tails get pecked often enough then the feather follicle is damaged and will probably never grow back properly.*

▶ *Roosting in the nestbox and therefore defecating in it: cover over the nestbox entrance during the afternoon in the hope that the offender will decide to roost instead. Make sure there are enough perches for all the hens to sit in comfort as one may have got bullied off the perches.*

DEATHS

It could happen that one day you find one of your birds dead in the hut. This may be because it is old or it may have had a heart attack (if the comb is purple on a hen or the head is dark on a turkey). If a bird is found dead and you have excluded other reasons than disease as the cause of death, such as vermin, it is sensible to have it autopsied in case there is something contagious that could affect the rest of your flock. Dispose of any carcases legally and never eat a bird found dead, only ones you have killed for that purpose.

How to cope with a broody hen

Refer to the laying chart on pages 4–5 for those breeds most likely to go broody. A broody hen has the instinct to sit upon eggs, keeping them warm and incubating them until they hatch. Hybrids have been selected so that broodiness is reduced but they can still sometimes go broody. You will find that one hen will insist on staying most of the time in the nestbox and may well try to peck you when you are collecting eggs. To check if she is really broody, gently slide your hand under her, palm up, and if she 'cuddles' your hand with her wings, then she is serious. The best way to reset her cycle is to construct a 'sin bin'. This could be as simple as a rabbit hutch with food and water available, but should really have a small mesh floor so that she does not want to think of a nest. It will take about a fortnight for this to be effective, but harden your heart and do not let her out before this time as she will go straight back to the nestbox and begin sitting again. After two weeks in the sin bin, integrate her carefully back into the flock (or place the sin bin in the henhouse if there is space) and she should begin to lay properly again quite quickly, therefore earning her keep. The next time she looks as though she is going broody put her in the sin bin straight away for a few days. This may be enough as the hormones are not at full power at the beginning of the broody cycle. (It is possible to have the hen injected by a vet with a hormone to reset the cycle, but that can become pricey.)

Selling eggs: the regulations

Since the beginning of 2004, any laying flock of over 350 birds must be registered with the Egg Marketing Inspectorate of DEFRA in order to sell eggs in the UK – readers in other countries should contact the relevant bodies. Most readers will be planning smaller flocks than this, but if they want to sell eggs other than to an end-user, such as restaurants, bed and breakfast enterprises, hotels, etc. the flock must be registered, no matter how small. There must be no cockerels in the flock to ensure that the eggs are not fertile – if stored in a warm place, a fertile egg could begin to develop. Once registered, only Class A eggs can be sold with the method of production printed on the package and on the eggs, plus the flock registration number and country of origin. Class A eggs must be clean, with no shell damage and sold by weight. All other eggs that have blemishes (internal or external) or are dirty must be consumed by the human producer. There are space regulations for free-range: birds must be stocked in the house at no more than nine per square metre (10.75 sq ft) and have access to an outside run with 4 sq m (43 sq ft) per bird. These regulations may seem somewhat draconian for the small producer, but are to ensure that traceability of genuine produce is maintained. Eggs from unregistered flocks of less than 350 birds presently sold in farmers' markets have had to be code stamped since July 2005 and those flocks registered. Contact your regional Egg Marketing Inspectorate at DEFRA for further details.

What to do when you want to go on holiday

A day out can be easily covered by checking the birds before you leave and if you will not be back before dark, asking a friend to close the pophole or fit an automatic closer. It is not acceptable to leave poultry for a weekend without them being checked, even if food and water is provided, so provision must be made for someone to do this.

If you want to go on holiday, someone will need to look at your birds at least once a day, feed and water them, collect the eggs and open and shut the pophole. By now you will hopefully have made fast friends with your neighbours and converted them to egg production as well, but livestock does tend to be rather tying if you want to go away. It may be that you can arrange for your neighbour to look after your birds in exchange for the eggs, or they may already be keen and helping you out. It is very unlikely that you will be able to board your poultry as you would a dog or cat, but there are various house- and pet-sitting services that may be worth contacting.

Breeding your own stock

TECHNICAL TERMS

Batch: the number of eggs set weekly in an incubator
Breeding pen: a cock and hens of one breed, selected for a
*   good Standard*

Brooder: electrically heated area for young chicks
Broody hen: a hen staying on the nest and incubating eggs
Clucker: a hen already broody
Dayolds: chicks less than 24 hours old
Electric hen: narrow insulated box on adjustable legs with
*   a thermostatically controlled element and a soft under*
*   material so the chicks can press up against it*
Embryo: developing bird in the egg
Fertile eggs: a cockerel must mate with the hens
Hatcher: separate thermostatically controlled insulated box
*   where eggs are placed two days before hatch date, cleaned*
*   between hatches*
Heat lamp: preferably ceramic for safety
Incubation: the time for each species to develop and hatch in the
*   optimum temperature; chickens take 21 days*
Incubator: electrically powered, thermostatically controlled,
*   insulated box with or without automatic turning, in which*
*   eggs are incubated*
Pot eggs: pottery or wooden eggs put in the nestbox to encourage
*   broodiness*
Selective breeding: only breeding from those which conform to
*   the Standard*
Set: to put eggs under a broody or in an incubator in order to
*   hatch them*
Sitting: a clutch of fertile eggs

All birds naturally breed in the spring as the length of
daylight stimulates their hormones and warmer weather
and insects give their young the best start. For the small
poultry keeper there is a choice of two methods of hatching
chicks – natural or artificial. The best idea is to gain
experience and confidence in both. Hen eggs take 21 days
to hatch, some very small bantam eggs may be a day early
and really large poultry such as Cochins may take a day extra.
Remember to feed a breeder ration to the adults for four weeks
before you want to set the eggs to increase both fertility and
hatchability and make sure only good-sized, normal-shaped
eggs are set.

If you do not have a cockerel, you will have to buy in fertile eggs. Ask if the parent stock has been fed on a breeder ration as this will increase the number of chicks obtained. A fertile egg remains dormant until it is placed at the correct temperature to develop. You may have the opportunity to buy in dayolds – try to make sure these have been sexed as it will remove the problem of cockerels later on. It will cost you more to raise hens than buying them in due to the economy of scale that rearers have, but the enjoyment will probably outweigh this.

NATURAL HATCHING

Insight
Natural hatching under a broody hen is the ideal way to raise a few chicks and children love the experience.

Natural hatching is essentially dependent on having a broody or broodies at the same time as the eggs you want to set. Silkie crosses make the best broodies; either x Wyandotte or x Sussex but most bantams will go broody. A small pen of these can be bred alongside the purebreds or layers. Successful hatching is achieved with attention to detail and keeping it simple. Nowadays, the bank of wooden sitting boxes has mostly given way to the disposable cardboard carton for hygiene reasons, but make sure it is in a fox-proof area, in a quiet spot away from other stock. The individual broody boxes should be large enough for the species and lined with short straw or woodshavings (hay produces harmful moulds), dusted with a pyrethrum-based insecticide, plus have good ventilation near the top. Broody birds of any species that sit beside each other spend their whole time stealing each other's eggs and generally ruining a hatch. The broody hen is best gently put into a dark cardboard box with straw and slowly taken to the broody box in the dark to keep her sitting and left for a day or so on a few unimportant or pot eggs to ensure she is still serious and not upset by the move. Some lighter breeds are best left where they decide to sit as if they are moved they will go off being broody. When a hen is broody, a bare patch of skin on her breast (brood patch) is in contact with the eggs to maintain temperature and humidity.

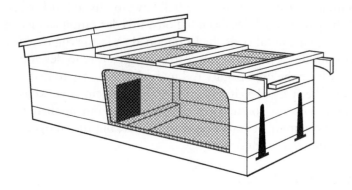

Figure 1.32 A broody coop and run.

When your broody has proved herself steady, put the eggs you want to hatch under her very gently, preferably at night, removing the others. Set an odd number of eggs as these fit better into a circle. If you want to set more than one broody at a time, either make sure that you set the eggs the same day so that they all hatch together, or keep the broodies out of sight and sound of each other as the noise of cheeping will make the other broody get off her eggs and help with the hatching ones if hers are not cheeping. The broody hen should be taken off the nest once each day to feed, drink and defecate. Roughly the same time each day makes for a quieter bird as she is a creature of habit.

The hen should not be disturbed for two days before the eggs are due to hatch and a chick drinker (small, so they do not drown) and chick crumbs should be left within her reach. The hatch may take three days to complete, and the early chicks need to be able to feed. Any eggs left after three days should be gently shaken beside your ear: if they rattle they are not fertile but be careful they do not explode! Try not to disturb the hen while the hatch is on, tempting though it is to see how many have hatched, as she needs to bond with her chicks and turn her sitting instinct into the more aggressive and protective maternal instinct, but gently removing empty shells in the dark will mean the other eggs can hatch with more space.

Should you wish to amalgamate broods under one broody, the hen will know after 24 hours which chicks are her responsibility and

then she will attack and kill any 'intruders' – this comes with the
territory, so knowing the behaviour will help prevent disasters. There
are some broodies who will gladly welcome any chicks, but they are
unusual. I once had a superb Silkie who, if placed within the sound
of chicks for 12 hours would go broody – a wonderful character.

If you buy in dayolds, give them a drink by dipping their beaks in
tepid water, then place them in a smallish cardboard box so they
stay warm and put them within sound of the broody hen for about
an hour. Then take them out, concealing them in your palm and
place them gently under her, removing the (false) eggs at the same
time. This is best done in the dark, but it will depend on what time
of day you get the chicks. Remember you have only 24 hours if you
wish to add more: she cannot count but she has colour vision and
can tell the difference between chicks.

ARTIFICIAL HATCHING

This is the use of an incubator to hatch eggs. Many poultry keepers
regularly use small incubators, the advantage being that incubation
conditions are instantly available at the flick of a switch. It saves the
extra space of pens for broodies and takes very little electricity to
run. Technical improvements have greatly increased efficiency, but
best results will be obtained with eggs that are between 24 hours
and seven days old and which have been stored in a cool (10° C or
50° F) place and turned daily. Mark the eggs in pencil with the date
and breed and keep records. Any dirt on the eggs can be scraped off
with a dry potscraper, the ideal being to have clean eggs in the first
place. If eggs do have to be washed, use water warmer than they
are to ensure that the membrane under the shell expands, keeping
bacteria out (cold water makes it shrink, drawing bacteria in) plus
a poultry disinfectant such as Virkon. The same disinfectant can
be used with safety to clean out incubators after a hatch. This is
most important for the success of future hatches as the bacteria and
debris produced by a hatch is phenomenal.

Follow the manufacturer's instructions for an incubator and run it
for a few days, checking the temperature with another thermometer.
Do not add any water – it seems to be a common misconception

that in the UK water needs to be added during the incubation process. Try to site the incubator in a place that does not vary much in average temperature such as a spare bedroom, but avoid direct sunlight. During the incubation process the eggs must be turned in order for the embryo to develop normally (the broody does this by instinct). If turning by hand do so at least twice a day and turn the eggs end-over-end so that the chalazae (strings that hold the yolk stable) do not wind up, potentially damaging the embryo, or mark the eggs and turn them from one side then back to the other, not continuously in the same direction. If the incubator is an automatic turning one, turn off the mechanism two days before the hatch date, or stop turning them by hand at that time. The best system is to have a separate hatcher so that eggs can be set weekly and not actually hatch in the incubator, therefore keeping it cleaner.

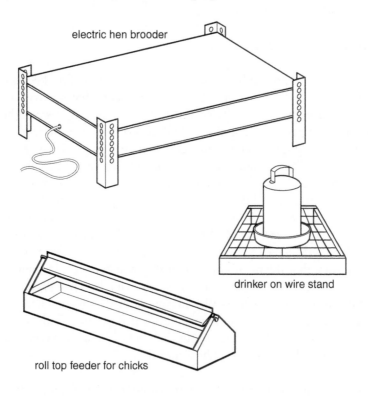

electric hen brooder

drinker on wire stand

roll top feeder for chicks

Figure 1.33 An electric hen brooder, a drinker on a wire stand and a roll top feeder for chicks.

Eggs should be moved to the hatcher two days before they are due to hatch. A little hot water can be added in the base of the hatcher when the eggs start to pip (the diamond-shaped start of the shell breaking) to keep the shell membrane moist. The chick pecks its way out of the broad end of the egg by means of the egg tooth which is on the end of its top beak. The egg tooth falls off soon after hatching. Chicks may take two days to hatch or they may all hatch at once. The latter is better, but not always possible. Most small incubators have a window in so that you do not have to take off the top to see inside. If you do not have a separate hatcher, it is better to fill (or part fill) an incubator, hatch the eggs, clean it out and start again. This avoids the build-up of harmful bacteria that can adversely affect the hatch.

CANDLING

In order to make best use of incubator space (and broody hens for that matter) the eggs can be candled after seven days' incubation. This involves holding a bright torch to the broad end of each egg in a darkened room. Obviously white-shelled eggs will be easier to see into than dark brown eggs. If the egg is infertile you will be able to see just the shadow of the yolk. Rotate the egg slightly to make this move within it. If fertile, a spider-shape of blood vessels will be seen with the heart beating in the middle. If there is a ring of blood vessels with none in the centre, the germ has died for some reason. The infertile eggs can be removed and fresh ones added if you are going to use a separate hatcher. If you are not sure, wait until the next candling session in a week when fertile eggs will be dark near the pointed end and the airspace will be much larger with a sharp dividing line between the two. If the division is fuzzy, or if there is only a small dark area, the embryo is likely to have died. The airsac gradually gets larger as hatching date approaches and sometimes the chick can be seen bobbing away from the candling light.

Another problem with hatching in the same incubator as other eggs that are still growing is that to get a good hatch, the humidity has to be raised and may well drown the developing eggs. Each egg should lose 13 per cent of its weight during incubation and this weight loss is mostly moisture so if there is too much moisture in the incubator, the correct amount of weight cannot be lost and the embryo either drowns or is so large that it is unable to turn and pip its way out of the shell. Most incubators in the UK do not need any water during incubation until two days before the expected hatch. People who hatch individually expensive birds such as raptors and parrots regularly weigh their eggs and have a series of incubators at different humidities so that the egg can be moved to another if the weight loss is not correct.

When cleaning the incubator or hatcher between hatches, use recommended disinfectants, because bleach-based ones, such as Milton, can damage any metal in an incubator. It is best to use Virkon, F10 or egg sanitizers. If you wish to set eggs on a weekly basis, then a separate hatcher is the only sensible solution as the huge amount of bacteria released on hatching is likely to infect the other eggs still incubating. The hatcher need not be very sophisticated, but needs a suitable thermostat, sufficient air flow to keep oxygen levels up and remove carbon dioxide, plus somewhere to put water to increase humidity so that the membrane inside the shell does not dry out. Get your hatcher set up at least a day before you think you need it and put warm water in the base so that the temperature is not reduced. Some people transfer chicken eggs on the 18th day; I prefer to candle the eggs and transfer only those looking viable (if you are unsure, put them in anyway as some breeds such as Cochins take another half day before hatching). Then do not open the hatcher until the chicks have hatched as you will lose all the valuable built-up humidity. Hatchers with a window will let you see what is going on.

Take the chicks out when they are dry and, keeping them warm and dark, transfer them to their rearing quarters and dip their beaks in tepid water.

REARING

Dayold to six to eight weeks
With modern equipment, rearing chickens is a relatively easy process for the small poultry keeper to master. If you have a

broody hen to do it for you, then all you will need to provide are chick crumbs, water and shelter against wind, rain and sun, preferably with a wired-over run so that magpies and crows cannot take the chicks or steal the food. Chick crumbs need to be in a container that the hen can neither tip over nor scratch them out of. Water needs to be in a container such that the chicks cannot drown. Grain feed should be left for the hen out of reach of the chicks. She may break the grain into small pieces for them. Leave the hen with the chicks for about four to five weeks and then take her away. Do not take the chicks away as it will unsettle them or set them back and they need all the encouragement they can get. They can be transferred to a larger house and/or run when they are about eight weeks old.

Incubated chicks need a heat lamp to keep them warm, preferably one with an infrared ceramic bulb so that they have heat and not light. The wattage will depend on the number of chicks with a 100-watt bulb being sufficient for a few chicks and a 250 watt bulb needed for 50 chicks. The heat without light avoids feather pecking as they then have natural light and darkness to maximize body and feather growth. Site the heat lamp in a draught-free place with a generous covering of shavings or newspaper on the floor or make a circle using a 2.4 m (8′) length of hardboard about 45 cm (18″) high around it. You can also use a large rectangular cardboard box and change this for each hatch. It needs to be rectangular so that the lamp is at one end and the chicks can regulate their own temperature by moving away from the lamp. Turn the heat lamp on two days before the chicks are due to hatch. It should be far enough off the shavings so that the temperature under it is about 39° C (102° F). If the chicks are too hot they will scatter to the edges, panting. If they are too cold they will huddle in the middle, cheeping loudly. The ideal is to have a small empty circle just under the lamp. Transfer the chicks from the incubator when they have dried and fluffed up. Dip their beaks in the drown-proof drinker (only use tepid water) and place them under the lamp. If they are thirsty and you give them cold water, the shock can kill them.

Provide chick crumbs a short distance away from the lamp in a container that the chicks cannot scratch the feed out of. They can stay in this area either until they outgrow it or until they are weaned off the heat lamp, at about six weeks. The lamp can be gradually raised and then turned off in the middle of the day if it is hot outside, not forgetting to put it back on at night. The chicks should be well feathered by this stage and able to keep themselves warm, with the lighter breeds feathering up quicker than the heavier ones. If you have used a light heat lamp, try leaving this off at night if it is reasonably warm outside when the birds are about four weeks old so that they get used to the darkness, otherwise they could panic and smother if you put them in a dark hut.

Electric hen brooders are good in that the chicks can mimic their behaviour with a broody hen by pressing their backs to the warmth, but check that the temperature is high enough underneath – the best ones have adjustable legs to cater for either large fowl or bantams.

Daily routine for artificial rearing
▶ *Add fresh shavings or take out the top layer of newspaper (if this needs to be done more than once a day the area is not large enough for the number of chicks).*
▶ *Clean the drinker and give fresh tepid water.*
▶ *Give fresh food.*
▶ *Check the height of the lamp and the comfort of the chicks.*
▶ *Handle the chicks so that they get used to it and tame down.*

Eight weeks onwards
People argue over when (and sometimes if at all) chicks should be given perches. As long as the perches are at least 5 cm (2″) wide there should not be a problem of bent breastbones from perches that are too narrow, and the lighter breeds certainly like perches. The breastbone begins as cartilage and is therefore softer than bone and gradually converts to bone as the bird matures, so with narrow perches early in life, it can bend and deform. A rearing house needs to be large enough so that all the chicks can shelter in it if the weather is bad with space for a feeder and drinker.

If an adult house is being used, block off the nestboxes with cardboard, as roosting in a nestbox is a habit almost impossible to break later, leading to dirty and partly incubated eggs. Young stock should be under observation all the time of development. Those with obvious physical defects should be removed. This will leave more room with clean houses and runs for the others to develop satisfactorily. Take precautions (by putting cardboard to round off corners) when moving stock to new houses so that they do not huddle in corners and smother. Continue to feed best quality rations.

Chicks should be offered chick crumbs of 20–22 per cent protein. Some brands contain a coccidiostat (a chemical to inhibit the number of coccidia (a potentially lethal parasite) in the gut) but these are being gradually phased out as the European Union (EU) wishes to reduce the chemicals in livestock feed. There is a vaccine called Paracox that is given by mouth to dayold chicks, which is very successful and immunizes them against this potentially fatal parasite. See page 212 for more details. Chick crumbs should be fed *ad lib* in a container with a series of small openings or a swivel top to avoid waste and contamination. There should be enough trough space for most of the chicks to feed at one time to avoid bullying. At about six weeks introduce growers' pellets over a few days and offer a little grit. When the birds reach about 18 weeks they can be changed, gradually, to a layers' ration of 16 per cent protein and a little whole wheat, plus mixed grit.

SEXING

Insight

Some breeds can be sexed by colour at dayold such as Welsummers (females have better defined markings) or Marans (male has a larger white spot on top of his head) but this is only about 80 per cent accurate.

Vent-sexing is not really an option for the small breeder as there are no visible reproductive parts and professional chick sexers spend some five years learning how to do it. Some colours of breeds can be sexed as they get their next plumage, for instance the Silver Grey Dorking male has a black breast from about seven weeks. Female chicks tend to grow their tail and wing feathers before the males during the first three weeks. Do not trust the comb development for sexing as the development is affected by many different factors and breeds differ in comb style and size.

The shape of chick feathers begins with both sexes' being rounded. When chick feathers start to moult between 10 and 12 weeks, new, sharply pointed and shiny plumage can be noticed on the backs of the males between the shoulder blades. The females' feathers remain rounded all their lives. Breeds vary in their speed of development, but the Silkie cannot be sexed by the plumage: 14 weeks is the earliest they can be distinguished by differences in combs, crests and size, so don't be taken in by what looks like a pair and turns out to be an older and younger cockerel.

At about five months old it is possible to grade young stock for colour and markings and other breed characteristics. It is also time to move cockerels to a house of their own and to take stock and assess the worth of the season's crop, giving preferential treatment to those that look likely prize-winners or future breeders. Any males not good enough to keep can be fattened, see page 188.

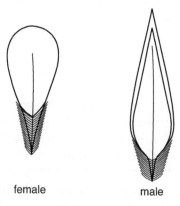

female   male

Figure 1.34 Female and male feather shapes after about ten weeks.

SELECTION FOR BREEDING PURE

Poultry give pleasure to thousands of breeders throughout the world. They provide relaxation for tired workers, help educate children in the pleasures and disciplines of livestock keeping and produce food for the owners in many a household where fresh eggs would otherwise be unknown. At a time when everyone tends to conform more and more, they allow the individual full scope for energy and creativity plus the conservation angle of keeping the pure and rare breeds going. The smallest of poultry keepers with only a tiny back garden can become a leading breeder. To do so, there are some basic principles to observe.

Selection

If the new poultry keeper intends to concentrate on serious conservation breeding and/or exhibiting it will be necessary to select parent stock very carefully:

▶ **Health:** *poultry that have suffered from any disease that has necessitated severe treatment should not be used for breeding. Good health is visible, it can be seen as well as felt when breeding stock is being selected. The feathers will be sleek and well furnished to form a protective covering during bad weather or hot sun. There will be a healthy glow about the head with no discoloration or weakness apparent. In most varieties, faces and wattles will be bright red. Discoloration or darkening of the comb might indicate liver or heart trouble, or it might be obesity, so the correct diet is important. There should be no wheezing that indicates respiratory or heart trouble. The face should be open with a pleasant expression. Eyes should be bright and stand out well from the head, nostrils dry. Plumage should be normal for the breed with each feather wide, well made, whole and resilient. Tail furnishings should be plentiful according to the Standard.*

▶ **Conformation:** *according to the breed the shape of the bird is laid down in the Standard. Examine the legs and feet: bones should be sound, neat, toes straight and refined. No breed should have coarse shanks and thick scales. Texture will be* *shown in tight, well-fitting scales continuing down the toes. Any deformity such as bent toes, duck feet, crooked breast bone, wry tail or split wing should mean rejection as a breeder.*

▶ **Breed character:** *all the poultry that have passed the handling test for health and conformation should be scrutinized for breed character. Because points vary for each breed it is necessary at this stage to know how many points are awarded for each shape, size or colour according to the different breeds, especially for show stock.*

▶ **Colour:** *detailed official colour (and type) Standards for each breed are to be found in the* British Poultry Standards *(Blackwells), overseen by The Poultry Club. Every breed has a Standard to which it must conform and every prospective breeding bird must carry good points of breed character and colour to accord with its breed name.*

▶ **Head points:** *these are especially featured because few Standards exist that do not give a fair share of points to formation of comb, lobes and wattles. Close inspection is necessary. Breeds with small single combs will not readily show up defects in females but will be latent and recur in cockerels of the following generation. Thin, glossy skin is not wanted in white-lobed breeds. It will soon yield to white in the face, a serious defect in showing and breeding.*

Mating up

After all the poultry have passed the above tests and are considered up to Standard and fit for breeding, the question arises of how many females should run with a male. With breeding, there is no hard and fast rule about this mating ratio. The breeder's target is quality rather than quantity dayold chicks. Thus many breeds, especially true bantams, are simply pair mated (one male to one female) or trio mated (one male to two females). This is very advantageous for pedigree records (see below). In the larger breeds they will be mated in pens of six or seven birds. The objective here is to get as many as possible from which to select those of high quality: when they are less robust the number of females that will run with one male is reduced. The surest way to progress is to try using birds that are similar in quality and possessing no bad faults.

Of course it is not always easy to come across birds for breeding that do not possess bad faults. Minor faults in one individual may be balanced by similar extra good points in the opposite sex. Having put the stock breeders together, eggs should be checked for shape, size, texture and colour. The better the egg, the more chance it has of producing a robust chick, if fertile. If egg shape and texture are neglected, the strain will gradually deteriorate until there are more weak eggs than there are good ones.

PEDIGREE RECORDS: THE POULTRY CLUB RINGING SCHEME

There is a limit to the amount of information needed for pedigree birds when details are either kept in the head or on the back of an envelope. Compulsory closed ringing is the norm in Europe for exhibition birds and is likely to be listed in the UK soon. The Poultry Club has had a voluntary ringing scheme for several years which is fairly well supported. Special prizes are given to rung birds at shows for encouragement, but it is essential for serious breeders to ring their birds so that they can keep track of family lines and traits. Rings are produced once a year, the colour changing each year, and each breeder has their own special number. A ring is placed on the leg of a chick at about four weeks of age, or at an age where the ring does not fall off and is reasonably easy to place on the leg. If a rung bird is sold, the ring number is transfered by the administrator of the Scheme. The ring number is allowed to be tattooed under the wing for extra security.

FERTILITY

If using a breeder ration, when egg laying commences under natural lighting conditions in the spring, fertile eggs may be expected within ten days of the male being introduced. If the male is already running with females it is possible that their eggs will be fertile from the first laying. If you have a different breed cock bird running with females, allow a fortnight for the correct bird to be fertile with those hens after removal of the other cock bird. It is not necessary for the male to copulate with each female daily. He can fertilize several eggs at one time if there is free access for the sperm to travel to the ovary. Some of the more fluffy breeds may need feathers removing from around their vents in order for successful mating to take place due to the fact that the birds do not have extrudable parts and merely touch vent to vent when mating in order to transfer the semen. You will find that cocks will have favourite hens and the feathers on the backs of these hens will be worn away. In order to prevent this if you want to show your birds, put the cock in with the hens for a few minutes every other day. It is not possible to determine from the actions of the birds how many eggs will be fertile, but if the stock is selected on the lines indicated, fed properly and allowed to settle down in the breeding pens, the percentage of fertile eggs should be quite high, according to breed.

CULLING

Culling is never easy: it doesn't necessarily mean killing a bird, but removing it from the breeding pen so that whatever fault it has is not perpetrated. Improvement in the standard of your stock is the goal and this includes not only superficial points but utility aspects as well.

There will, however, inevitably be cockerels that you do not want to keep and only a very, very few will be able to be sold for breeding, so before breeding any birds, this surplus needs to be considered. Should you wish to eat them yourself, that is fine, at least you know how they have lived and what they have been fed on.

On a small scale, the humane way to kill a chicken is by dislocating the neck. There are two sets of blood vessels in the neck and only by dislocating it can you disrupt blood and nerve supply to the brain and therefore first render the bird unconscious and then shortly afterwards, dead. Neck dislocation should only be carried out if immediate unconsciousness is induced without causing pain or suffering. Small numbers of birds on home premises can be killed by neck dislocation without prior stunning, although this may take two people with a large goose or turkey. (Stunning is usually done in slaughterhouses with one electrode on the overhead line where the birds are hung and the other in a tank of water where the birds' heads are dunked, thereby rendering them unconscious before they are bled.) It is the responsibility of the keeper to ensure that poultry are killed humanely. Dispose of carcases legally.

Either get an experienced poultry breeder to show you how to kill a bird humanely, or practise neck dislocation on a dead bird first.

MOTIVATION

When selection has been accurate, when mating up has given good, fertile eggs, when hatching has produced strong, healthy chicks, when rearing has brought those pictures in the books to life – this may be the time to seek comparison with other like stock. The route to follow is through the shows held under Poultry Club Rules. First the small local, next the more ambitious regionals, then to the big one where most of the Club shows are held – the National Championship Show, where every conceivable breed of poultry in the UK has classes of its own, usually held in mid-winter. Breed winners (and runners-up) are taken to Championship Row and judged by a separate judge, to establish Show Champion and all the other major awards. See Appendix 3, page 233.

SELLING SURPLUS STOCK

Ensure that birds are healthy, of the breed described and sexed. Advertisements may be placed in the country magazines mentioned earlier or in the local press. Be prepared for potential buyers asking lots of questions and wanting to see the parent stock. There are very few transport firms that will deliver poultry and these tend to be expensive, so the general practice is to make a trip yourself.

10 THINGS TO REMEMBER FOR CHICKENS

1 *Check local authority regulations.*

2 *Decide on the breed.*

3 *Get housing sorted before the chickens arrive.*

4 *Get feeders, drinkers and food before the chickens arrive.*

5 *Buy good stock from a reputable source.*

6 *Check the stock is healthy and vaccinated if necessary.*

7 *Keep out wild birds.*

8 *Collect eggs daily.*

9 *Check the chickens twice daily.*

10 *Shut them in at night.*

2

Ducks

In this chapter you will learn:
- *which breed to choose*
- *how best to look after your ducks*
- *how to breed them successfully.*

Which breed is best for you?

Domestic ducks are all descended from the wild mallard
(*Anas p. platyrhynchos*) except the Muscovy (*Cairina moschata*).
The lighter breed ducks certainly lay more eggs during the year,
but the heavy breed ones will lay in the spring and summer and
the variation in colour, shape and size makes for a more interesting
flock. The Muscovies and the bantam ducks are able to fly, so
provision must be made either to net them over or to clip flight
feathers (see page 48).

Do not keep ducks in the same area as chickens as they tend to be
rather messy and DEFRA (UK Department for Environment, Food
and Rural Affairs) advises not to keep waterfowl and chickens
together in order to lessen the risk of transfer of viruses and
other diseases.

Check with your local authority that there are no regulations
to prevent you keeping ducks and remember to inform your
neighbours, reassuring them that there will not be a sea of mud

Figure 2.1 Pekin duck: the origin of the meat breeds – note the upright carriage like a boat on its stern and the chubby face.

or smell. Female ducks can make a certain amount of noise, but not nearly as bad or as early as a cockerel. Remember, you do not need a drake for egg production and ducks are polygamous, mating with any female, with the exception of the Call ducks who need to be allowed to choose their own mate.

TECHNICAL TERMS

Bantam ducks: small ducks, used for exhibition or slug control
Bean: raised, hard, oval protruberance at tip of upper bill

Bill: beak
Breed Club: collection of enthusiasts for one breed
British Waterfowl Association: club for domestic and wild waterfowl
Broody: any duck wanting to incubate eggs
Call ducks: noisy, vocal very small ducks, used for exhibition
Caruncles: wart-like protruberances on head of male Muscovy
Clutch: a number of eggs laid by one duck until a day is missed
Coverts: small feathers on wing
Domestic Waterfowl Club: club for domestic waterfowl
Drake: male duck
Drake's tail: upward curled tail feathers in mature males
Duck: female duck
Duckling: young duck up to about six weeks
Eclipse: dull summer plumage of coloured drakes
Eyestripe: different coloured plumage through or around the eye
  in ducks
Grower: between six weeks and adult plumage (16 weeks)
Heavy ducks: dual purpose, eggs and meat
Incubation: keeping eggs at the correct temperature and humidity
  so they hatch
Incubation time: ducks take 28 days, Muscovies take 35 days
Keel: deep, pendant fold of skin and tissue
Lacing: a different colour line around each feather
Light ducks: the more active breeds
Lustre: sheen
Mallard: ancestor of all domestic ducks except Muscovy
Muscovy: South American perching duck
Neck ring: white ring of feathers either completely encircling the
  neck (closed) or open (at the back)
Pair: male and female
Poultry Club of Great Britain: responsible for Standards
  (www.poultryclub.org)
Primaries: large feathers on outside of wings
Secondaries: smaller feathers on middle of wings
Speculum: irridescent band of colour on secondaries in ducks
Sport: an unusual colour produced from pure breeds
Standards: published characteristics and plumage colour for each breed
Trio: male and two females

HYBRIDS VERSUS PURE BREEDS

Unlike chickens, there are very few commercial hybrid type ducks available. In fact, the pure light breeds are more than adequate for laying purposes. It is most important to make sure of a market for the eggs. They are supreme in baking, generally having a better yolk colour than hen eggs (via natural feeding) plus extra moisture – duck eggs are the secret ingredient of prize-winning and popular cakes and cooking.

Light breed ducks should lay well for about three years and then, for continuous production, they should be replaced either by buying in or breeding your own.

Insight

Should you want slug control in a vegetable garden, the smaller bantam or Call ducks are useful as they do not do much damage with their feet, but you may need to clip the feathers on one wing as these birds fly well. Indian Runners do not fly and are also good sluggers.

CLASSIFICATION OF DUCK BREEDS

Heavy
Aylesbury
Blue Swedish
Cayuga
Muscovy
Pekin
Rouen
Rouen Clair
Saxony

Light
Abacot Ranger
Bali
Buff Orpington
Campbell
Crested
Hook Bill
Indian Runner
Magpie
Welsh Harlequin

Bantam
Black East Indian
Call
Crested
Silver Appleyard Miniature
Silver Bantam

A GUIDE TO PURE BREED DUCK EXPECTED LAYING CAPABILITIES

Breed	Egg colour	Numbers per annum	Maturing	Type
Abacot Ranger	white	250	quick	light
Aylesbury	white	20	slow	heavy
Bali	white	200	quick	light
Black East Indian	dark grey	15	quick	bantam*
Blue Swedish	blue/green	150	slow	heavy
Call	white or blue/ green	30	quick	bantam*
Campbell (Khaki)	white	320	quick	light
Cayuga	dark grey	40	slow	heavy
Crested: large	white	150	quick	light
miniature	white	15	quick	bantam
Indian Runner	blue/green	295	quick	light
Magpie	blue/green	150	quick	light
Muscovy	white	50	slow	heavy*
Orpington	white	150	quick	light
Pekin	white	40	slow	heavy
Rouen Clair	white	80	slow	heavy
Rouen	white	30	slow	heavy
Saxony	white	150	slow	heavy
Silver Appleyard: large	white	150	slow	heavy*
miniature	white	40	quick	bantam
Silver bantam	white	40	quick	bantam
Welsh Harlequin	white	250	quick	light

* most likely to go broody

LIGHT BREEDS (NOT RARE)

Abacot Ranger
This was one of many breeds developed from (or crossed with) Indian Runners. Starting with 'sports' from Khaki Campbells, themselves originally the products of Runner crosses, Mr Oscar

Gray of Abacot Duck Ranch, near Colchester, mated their offspring to a white Indian Runner drake. The eventual results were 'light drakes of Khaki carriage and type with dark hoods, and white ducks with blue flight bars and fawn or grey hoods'. *Colour*: the drake is similar to the wild mallard with a green head, open white neck ring, claret breast, grey back and belly and black tail but with white fringing on the breast feathers, bill olive green. The duck has a fawn head and paler body streaked with brown, bill dark slate green. The speculum must be blue-green. Weights: drake 2.3–2.5 kg (5–5½ lb), duck 2.0–2.3 kg (4½–5 lb).

Buff Orpington

The Orpington Ducks were developed by William Cook of Kent, who also bred Orpington chickens in the late nineteenth century. The Buffs were standardized in 1910, followed by the Blue variety in 1926. The Standard Buff Orpington is an attractive but unstable colour form, standard parents producing three variations in colour of offspring. *Colour*: Buff, drake darker than the duck. Drake bill yellow ochre, duck bill dark brown. Weights: drake 2.2–3.4 kg (5–7½ lb), duck 2.2–3.2 kg (5–7 lb).

Crested

....
Insight
> Looking as though they are wearing hats, Crested ducks are portrayed in Dutch paintings in the 1600s.
....

Crested ducks are, again, not easy to breed for exhibition, but are comical characters and lay reasonably well. *Colours*: white is the most common, but any colour is permitted as long as the markings are symmetrical. Bill colour relates to body colour. Weights: drake 3.2 kg (7 lb), duck 2.7 kg (6 lb).

Indian Runner

Originally from Malaya, these odd-looking upright ducks are very active and good layers. They should resemble a walking stick or wine bottle on legs. They can be a little nervous and need handling from a very young age to overcome this. *Colours*: White (bill orange), Fawn and White (bill green-yellow), Black (bill black), Fawn (bill black), Chocolate (bill black), Trout (bill golden olive in drake, orange-brown in duck), Mallard (bill green in drake, dark orange in duck). Weights: drake 1.6–2.3 kg (3½ –5 lb), duck 1.4–2.0 kg (3–4½ lb).

Khaki Campbell

One of the first, and certainly the most successful, of the utility breeds designed in the twentieth century from the Indian Runner, the Khaki Campbell largely took over as the top egg laying duck. The wild mallard, the fawn-and-white Runner and the Rouen were used to create this utility breed in 1901 by Mrs Campbell of Uley, Gloucestershire, specifically as a laying duck, expecting about 300 eggs per year. The colour of farmyard mud, from which it gets the first part of its name, the Khaki Campbell proved to be exceedingly agile, very fertile and extremely prolific. There are other colours of Campbell but they do not lay as well, the White Campbell coming as a sport from the Khaki. *Colour*: Khaki (dark bill), White (orange bill), Dark (dark bill). Weights: drake 2.3–2.5 kg (5–5½ lb), duck in laying condition 2.0–2.3 kg (4½ –5 lb).

Magpie

This is an unusual duck breed named only according to its original plumage markings of black and white. It was developed by Rev Gower Williams and Mr Oliver Drake in the years following World War I. It is difficult to get the correct markings for exhibition, but the Magpie is a good layer. When breeding them, some can come out all white, so if you keep other white ducks such as White Campbell, make sure you can tell them apart – the Magpie tends to be longer in the neck and back than the Campbell and the egg colour is different. *Colours*: black (on head and back, white elsewhere), blue (on head and back, white elsewhere), chocolate (on head and back, white elsewhere). Bill yellow in all colours. Weights: drake 2.5–3.2 kg (5½ –7 lb), duck 2.0–2.7 kg (4½ –6 lb).

Welsh Harlequin

Originally bred by Group Captain Leslie Bonnet at the end of World War II, the breed owes its existence to a chance production of two 'mutations' or sports from a flock of Khaki Campbells in 1949. These 'Honey Campbells' were renamed 'Welsh Harlequins' when the Bonnet family moved to Wales. *Colour*: drake is a brown version of the mallard colouring, bill olive green. The duck is fawn with brown markings all over, bill dark slate green. The speculum must be bronze. Weights: drake 2.3–2.5 kg (5–5½ lb), duck 2.0–2.3 kg (4½ –5 lb).

Figure 2.2 Trout Indian Runners: the colour is called Trout from some similarity to fish markings, the drakes are the darker ones but have not coloured up yet as they are young.

HEAVY BREEDS

Aylesbury

The Aylesbury derives its name from the town of Aylesbury in Buckinghamshire, where it was bred as a white table duck in the eighteenth century and used in increasingly large numbers to supply the London market. The white plumage was valued for quilt filling and the pale skin contributed an attractive carcase. This light colouration is evident in the pink bill colour that continues to be a key characteristic of the breed. The Aylesbury was a leading exhibit in the first National Poultry Show held at the London Zoological Gardens in London in June 1845. This was the beginning of live poultry exhibitions, and it was the Victorian stress on size that led to the development of the modern Aylesbury with its pronounced keel and long, pink bill. It was standardized in 1865. Now considered rare. *Colour*: white with flesh-pink bill. (Commercial meat ducks are based on the Pekin, therefore off-white with yellow bill.) Weights: drake 4.5–5.4 kg (10–12 lb), duck 4.1–5 kg (9–11 lb).

Blue Swedish

Blue ducks emerged in northern Europe during the nineteenth century in various breeds. The Blue Swedish was developed near the Baltic shores of what is now modern Germany and Poland. It is a medium-sized, utility bird with colour genes that allow them to produce three types of offspring (blue, black and a pale silver or splashed form). The standardized blue is slate blue with darker lacing (line round each feather), a white bib and two white primaries. *Colour*: Blue. Bill in drake blue, in duck slate blue. Weights: drake 3.6 kg (8 lb), duck 3.2 kg (7 lb).

Cayuga

This breed takes its name from Lake Cayuga in New York State. Thought to be descended from the wild black duck (*Anas rubripes*), the Cayuga was recorded in North America between 1830 and 1850. It was first standardized in America in 1874 and in Britain in 1901. *Colour*: black with brilliant green lustre, bill black. Weights: drake 3.6 kg (8 lb), duck 3.2 kg (7 lb).

Muscovy

Insight

The only domestic duck not descended from the mallard, the wild Muscovy is a native of South and Central America where it was being domesticated long before Columbus arrived in 1492.

The Muscovy is a member of the perching duck family and these are heavy-bodied birds with relatively short legs that give them a horizontal carriage. They have broad wings with bony knobs at the bends and the males show a slight swelling of the forehead during the breeding season. They have a forecrown crest that is erected in display, wart-like 'caruncles', especially in the case of the males, and webbed feet equally suited to perching as swimming with long and strong claws. They are capable of flight. *Colours*: black, blue, chocolate, lavender, black magpie (black plus white), blue magpie, chocolate magpie, lavender magpie. Bill colour variable and linked to plumage colour. Weights: drake 4.5–6.3 kg (10–14 lb), duck 2.3–3.2 kg (5–7 lb). It is a characteristic of the breed for the drake to be about twice the size of the duck.

Pekin

The Pekin was imported from China into Great Britain and America between 1872 and 1874. Crossed with other breeds, it had a large impact on the commercial table bird market. The American strain continues to show a less than upright carriage whilst the British and European Pekins retain the chubby 'boat standing on its stern' characteristics. This is a variety that continues to thrive as an exhibition breed and also as basic stock for commercial breeding. In addition, it provided genetic material for several modern breeds, including the Saxony. *Colour*: cream, bill yellow. Weights: drake 4.1 kg (9 lb), duck 3.6 kg (8 lb).

Rouen

The Rouen is a very large domestic duck with plumage colour and markings that resemble those of the wild mallard. Originally produced in Normandy, in the Rouen area, it was imported into southern England some time in the eighteenth century and its size, shape and colouring were further developed. This was one of the first duck breeds standardized in Great Britain (in the original Standards of 1865), and is now valued more for its size and plumage markings in the exhibition pen than as a commercial meat bird, despite the look of the deep keel which in fact is skin, not meat. It has also been used as basic stock for breeding many of the duck breeds developed in the twentieth century. *Colour*: the drake is similar to the wild mallard with green head, white neck ring, claret breast, black back and rump, grey flanks and abdomen and green-yellow bill. The duck is a rich golden brown on each feather with darker brown graining on head, then distinctly pencilled from breast to flank and stern, the markings very dark brown to black, the black pencilling on the rump having a green lustre. Large feathers ideally show double or triple pencilling (chevrons). Bill orange with black saddle. Speculum blue. Weights: drake 4.5–5.4 kg (10–12 lb), duck 4.1–5 kg (9–11 lb).

Saxony
Developed mainly from the Rouen, German Pekin and Blue Pommern around 1930, the Saxony was bred in Germany and came to Great Britain in 1982. It is a pretty, dual-purpose breed, laying reasonably well and also producing a good carcase. *Colour*: drake has a blue head, closed white neck ring, claret breast, blue back and oatmeal belly, bill yellow. The duck is deep apricot buff on the head with a dark line through the eye and creamy eyebrow, cream throat, paler buff back, and oatmeal wings, bill yellow. The speculum is blue. Weights: drake 3.6 kg (8 lb), duck 3.2 kg (7 lb).

Silver Appleyard
Originated by Reginald Appleyard in the mid-twentieth century, the Silver Appleyard was developed as a layer of 'lots of big white eggs' and as a white-skinned table bird. It is active and attractive. *Colour*: the drake has a green head with a white-flecked throat and faint silver eyestripes, closed white neck ring, claret fringed with white breast, dark grey back, grey belly, black tail and bill yellow-green. The duck is silver-white with markings of fawn flecked with brown, darker fawn eyestripe, fawn and brown back, silver belly, bill yellow. The speculum must be blue. Weights: drake 3.6–4.1 kg (8–9 lb), duck 3.2–3.6 (7–8 lb) kg.

BANTAM BREEDS (NOT RARE)

Black East Indian
First Standardized in Britain in 1865, the Black East Indian duck looks similar to the Cayuga. This bantam duck had been in the possession of the London Zoological Society since 1831. At this time it was known as the 'Buenos Aires' duck, but there seems to be no evidence that South America or the East Indies were the places of origin. It has been known as 'Labrador', 'Brazilian',

'Buenos Aires' and eventually 'Black East Indie', the former being perhaps the most appropriate geographically. *Colour*: black with beetle-green lustre. Bill black. The females can acquire white feathers with age. Weights: drake 0.9 kg (2 lb), duck 0.7–0.8 kg (1½–1¾ lb).

Silver Appleyard miniature
This is the same colour as the large version, but a third of the weight. Weights: drake 1.4 kg (3 lb), duck 1.1 kg (2½ lb).

Silver Bantam duck
Similar colouring to the Abacot Ranger. Weights: drake 0.9 kg (2 lb) duck 0.8 kg (1¾ lb).

CALL DUCKS

Insight
Named from the Dutch word *kooi*, meaning 'cage' or 'trap', decoy ducks have been used for centuries in Holland and Britain to lure wild ducks into complicated mesh structures.

The early Decoys were in fact little different from the wild birds, being decoys by training and performance rather than looks. Two key characteristics led to their modern development: they had to be highly vocal (hence 'call' ducks) to attract the wild ducks, and they benefited from being small and easy to carry. It was the Dutch Decoys that introduced the characteristics so prized in these birds. Modern Call ducks have genes that express dwarfism: small, shortened bodies and necks; large, round heads; short bills, big eyes and chubby cheeks. This has added significantly to the aesthetic appeal of the modern birds, but many of the early Calls were quite different from the exhibition birds of today. Less than a hundred years ago illustrations showed Calls with appreciably longer bodies and bills. There have been many colours developed, mainly for exhibition, but these diminutive ducks are great characters, good with children and take up little space. If you wish to breed them, let the ducks choose their own mate, which increases the success rate. *Colours*: Mallard, Blue Fawn, Apricot (linked series), Silver,

Blue Silver Apricot Silver (linked series), Mallard Dusky, Bibbed, Mallard Pied, Magpie, and the classic White, the model for all bathtime ducks. Weights: drake 0.6–0.7 kg (1¼ –1½ lb), duck 0.5–0.6 kg (1–1¼ lb).

Buying

Study the health points and follow the biosecurity guidelines below so that you know what to look for when going to buy birds. Reject any that do not come up to scratch: don't buy them because you feel sorry for them – they will be nothing but trouble.

BIOSECURITY FOR DUCKS

▶ *Isolate new stock for two to three weeks.*
▶ *After exposure at an exhibition isolate birds for seven days.*
▶ *Change clothes and wash boots before and after visiting other breeders.*
▶ *Change clothes and wash boots before and after attending a sale.*
▶ *Keep fresh disinfectant at the entrance to waterfowl areas for dipping footwear.*
▶ *Disinfect crates before and after use, especially if lent to others. However, it is preferable not to share equipment.*
▶ *Disinfect vehicles that have been on waterfowl premises but avoid taking vehicles onto other premises.*
▶ *Wash hands before and after handling ducks.*
▶ *Comply with any import/export regulations/guidelines.*

These are common-sense measures that can easily be incorporated into a daily routine.

POSITIVE SIGNS OF HEALTH IN DUCKS

▶ *Dry nostrils.*
▶ *Bright eyes (colour varies with breed), no soreness.*

- *Clean, shiny feathers (all present).*
- *Good weight and musculature for age.*
- *Clean vent feathers with no smell.*
- *Straight toes and undamaged webs.*
- *The bird is alert and active with no sign of lameness.*

AGE OF ACQUISITION

Ducks mature faster than hens and are sold at 12–14 weeks, already sexed, as they should start to lay from about 16 weeks. Or they are sold to the more experienced breeder at dayold, sexed (see page 113). Most breeders of exhibition stock will only sell pairs or trios, i.e. including a drake, so try to negotiate for females not quite good enough for the show pen, for instance, remembering that the heavier breeds will lay less in any case.

WHERE TO BUY

Check all stock using the health signs above before purchase.

- *From advertisements in smallholding magazines such as* Country Smallholding, Smallholder, Fancy Fowl, Practical Poultry *and* British Waterfowl Association Yearbook.
- *From private breeders who exhibit at waterfowl shows.*
- *From private breeders. Ask to see the parent stock.*
- *At small sales. Talk to the breeder.*

WHERE TO AVOID

- *Large sales as prices can escalate and there may not be any history with the birds such as age, health status and how they have been reared.*
- *Adverts in local newspapers may be genuine or dodgy – if it sounds good, go to see the birds before purchase.*
- *Car boot sales and the internet.*

84

Handling ducks

If ducks are in a house and run then driving them into the house will be the best method to catch them. Then corner the duck you want, restraining it loosely around the neck before sliding your other hand underneath from the front, palm up, and clasping its legs between your fingers. Ducks wriggle, so hold the legs firmly. Transfer the duck to your forearm, with the other hand on its back and its head pointing behind you with the mucky end pointing away from you – ducks tend to projectile defecate when picked up, so you really do not want their mess in your pocket.

If you are getting a duck out of a box or crate, loosely restrain it around the neck, then slide your hand in under the bird from the front, palm up, and clasp the legs firmly, then transfer as above.

If ducks are free-range or have a large pond, they will soon learn that being on the water is the safest place and you will be unable to catch them. Always try to be devious first and feed them away from the water so that they can then be driven into a hut or run. Once in the run, unless they have been used to being handled from dayold, it is best to catch them with a fishing landing net, then transfer them to an arm as above.

Be especially careful to put ducks down gently after handling as they can damage a leg quite easily.

Insight
> Ducks rarely bite (and it is only a little nip if they do), but do beware of Muscovy claws as not only are they sharp, the ducks are extremely strong for their size.

It is permissible to pick a duck up with fingers around the base of both spread wings in one hand, middle finger pointing down the duck's back, for a very short transfer distance – useful with small, wriggling wildfowl whose legs are vulnerable to damage.

Figure 2.3 Aylesbury: this is a champion exhibition bird, quite different from the Pekin.

Start-up costs and other considerations

It used to be cheaper to start up with ducks rather than chickens, but with the threat of avian influenza (AI) and the directive to exclude wild birds from feeders and drinkers and to be able to keep all domestic poultry indoors if necessary, provision will have to be made using either a covered run or netting over an area.

START-UP COSTS

Duck house and run (e.g. for 6 ducks)	£270	($405)
6 laying ducks @ £10	£60	($90)
Drinker and feeder	£30	($45)
Plastic dustbin for feed storage	£5	($7.50)
Plastic movable pond	£15	($22.50)
Total	£380	($570)

86

Feed: allow 85 g of layers' pellets per bird per day,
6 ducks will eat 25 kg (one bag) in 7 weeks (8 × £6)   £48   ($72)
Wheat 150 kg (@ £4 per 25 kg)   £24   ($36)
Grit, shavings or straw   £20   ($30)
                                                        ‾‾‾‾
Total   £92   ($138)

It is easy to waste feed, so get the proper feeding equipment,
whether galvanized or plastic.

INCOME/BENEFITS

▶ *Maximum production from laying ducks such as a Khaki
  Campbells is 320 eggs per year, about six eggs per week per
  duck, total 36 eggs per week. If the family eats 18 eggs per
  week, the remainder can be sold for £2–2.50 ($4) per dozen,
  giving £156–195 ($234–293) income, which can go towards
  covering the cost of the feed.*
▶ *If ducks are kept in a hut then the manure can be used in your
  own garden or allotment, or sold. It does not contain quite
  as much nitrogen as chicken manure, but still needs to be
  composted before using.*
▶ *Slug control in a vegetable garden or allotment.*
▶ *Hours of observation and enjoyment!*

Housing

TECHNICAL TERMS

Apex roof: two slopes
*Fold unit: movable self-contained house and run, may or may not
    have wheels*
Free-range: access to grass in daylight
Hut, house: duck house

Litter: dry and friable substrate on the floor
Nestbox: to lay eggs in
Pent roof: one slope
Pond: preferably movable and with a ballvalve
Run or pen: fenced exercise area, usually grassed
Shavings: livestock woodshavings for litter, also to line nestbox
Skids: used to move larger huts
Straw: usually wheat straw as barley straw is too soft
Ventilation: must be at roof level and above heads of birds
Window: replace any glass with wire mesh

Housing is used by ducks for sleeping, laying and shelter. The welfare of the birds is entirely in your hands and certain principles should therefore be observed.

SPACE

Housing is needed for waterfowl at night for safety from predators. The floor area should be a minimum of 0.75 × 0.75 m (30 × 30″) for each light duck upwards. Ideally, they will all be in a fox-proof enclosure so will not need secure housing, especially as waterfowl see well in the dark and really do not like going into huts, except to lay. Make sure the huts have fairly high interiors so the birds feel less claustrophobic if they have to be shut in at night. A hut with a large entrance door, one that perhaps drops down, will encourage the birds to go in. Bantam ducks are going to need half the space of larger ones. If the area is fox-proof, then a very simple shelter for laying in is all that is needed.

WATER

Insight
A child's plastic sandpit makes a good movable pond, but build a ramp so the ducks can get in and out easily.

Bantam ducks do well in small enclosures with a pond or water container that is easy to clean and positioned either on grass or gravel.

Water for larger ducks is ideally provided in movable ponds that have a ballvalve, thus keeping the level of water constant plus avoiding the muddy patches that always appear due to their habit of dabbling. A pond that can be emptied at least once a week is adequate. If a natural or dug-out pond is very large, put small mesh netting around all the banks to prevent the waterfowl from digging them out, as this is one of their favourite activities. Do not think for one minute that plants will survive in a pond with ducks, unless it is a large pond and has only one pair of bantam ducks. Pea gravel around a fixed pond will help to keep the area cleaner and this can be hosed down. No matter how disciplined you think you are, waterfowl acquisition is addictive, so allow for serious expansion when planning your enclosures.

It is important that ducks have a wide enough drinker so they can dunk their whole head in order to keep their eyes clean.

Figure 2.4 A movable duck pond with ballvalve.

FEEDING

Use commercial feed for the correct species and age of bird and put this in vermin-proof hoppers. Hanging feeders with a coiled spring at the base works well for pellets with ducks, and wheat can be put in shallow troughs that are filled with water so that crows and rooks are not attracted to the feed. Ideally net it over.

VENTILATION

It is vital to provide good ventilation as the thick feathering of waterfowl is all they need for protection from cold. Housing is used to provide protection from predators rather than weather protection. Wire mesh windows or doors are best to allow good air circulation, but don't make the windows too large as moonlit nights will spook the ducks.

NESTBOXES

Ducks are less agile than chickens and will not readily use the sort of outside nestbox available on most henhouses. Therefore duck huts tend to be simpler and nesting areas for waterfowl can consist of some overhead protection and wheat straw on the ground. A simple triangular hut with no base and one side open is a good laying area for outside waterfowl as it affords some protection from aerial predators. Inside a duck hut, create nestboxes by leaning hardboard at about 30° against one wall to make a dark and protected area. Letting ducks out after 9 am will ensure that 99 per cent of the eggs are laid in the hut, as they lay at the same time every day (see Chapter 1, Breeding your own stock, Figure 1.28 and page 53 onwards).

PERCHES

Insight
Waterfowl sleep on the ground and do not want a perch, except for Muscovies which do like to perch.

POPHOLE

This is a low door to enable the ducks to go in and out of the house at will in daylight. The most practical design has a vertical sliding cover that is closed at night to prevent fox damage. The horizontal sliding popholes quickly get bunged up with muck and dirt and are difficult to close. Light sensitive or timed gadgets are available that will close the vertical pophole if you have to be away at dusk. Most commercial duck houses have fairly high and wide popholes as ducks do not like to bend to go into a hut and they can push and shove on the way out in the morning. If the pophole is off the ground, make a ramp for the ducks as they cannot easily jump over a step.

MATERIALS

Timber should be used for the frame, which can then be clad with tongue and groove, shiplap or good quality plyboard. If the timber is pressure treated by tanalizing or protimizing it will last longer without rotting. Recycled plastic has recently come on the market as a duck housing material with a sliding roof for good people access and a good wide pophole. It is rot-proof, light, easy to clean and less likely to harbour parasites, but check that there is enough ventilation for early summer mornings when the sun is hot and you are waiting to let them out in order to get the eggs.

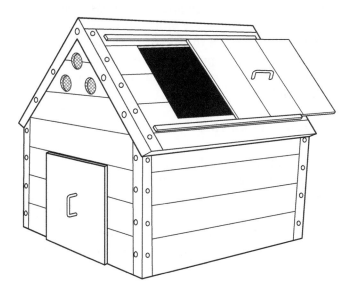

Figure 2.5 A recycled plastic duck or goose house.

LITTER

Wheat straw or woodshavings on the floor and for nesting areas are the best materials. Do not use hay due to the mould organisms present in it.

FLOOR

The floor can be solid, slatted or mesh. Slats should be 3.2 cm (1¼″) across with a 2.5 cm (1″) gap between, mesh should be 2.5 sq cm (0.38 sq in). If slats or mesh are used, make sure the house is not off the ground otherwise it will be too draughty. Slats or mesh make for better drainage and should be covered with straw. If the hut has a solid floor, raise the house off the ground about 20 cm (8″) to deter rats but you will then need a ramp for the ducks to get back in.

CLEANING

Weekly cleaning is best, replacing litter in all areas. The best disinfectant that is not toxic to the birds is Virkon. This is a DEFRA approved disinfectant that destroys all the bacteria, viruses and fungi harmful to poultry. Remember to replace the litter in the nesting area and move all housing on a regular basis to help with hygiene.

PEN, RUN OR FREE-RANGE?

With the new regulations (see Appendix 4), provision must be made to at least net over a duck run in order to exclude wild birds. If ducks are kept in a covered run, this can be moved on a regular basis so that the ground does not sour or become waterlogged. If the ducks are to free-range, keep the feeder inside the house (again, to prevent wild bird access), as they will forage and get quite a bit of their diet from insects and herbage.

Beware poisonous plants such as laburnum, laurel and nightshade, but if you have children you won't have these in your garden anyway.

Daffodil bulbs are toxic, so be careful of these, although most poisonous plants taste horrible to ducks. Unless the covered run is a large area, don't attempt to plant shrubs inside it as the ducks will soon dig around the roots of these and most likely kill them. Clematis, honeysuckle, berberis, pyracantha or firs can be grown on the outside of the run both for shelter and to enhance the area.

If you are gardening, the ducks will love to help you catch slugs, worms and insects.

If you want to weed an allotment, use a fold unit which is a house and run combined that can be moved to a fresh piece of ground as soon as the ducks have done their job, possibly daily, which means any droppings can be incorporated immediately as there will only be a few. If the ducks are contained within the fold unit (feeder and drinker hang in the run part) they will efficiently weed, de-slug and manure an area of your choice and leave your precious vegetables alone, plus being protected from the fox. A large drinker, enabling the ducks to immerse their heads, is important, in order to keep their eyes clean.

BUY OR MAKE?

If housing is bought from a reputable manufacturer and meets all the basic principles then that may be the quickest and easiest method of housing your birds. If you wish to make housing yourself, keep to the basic principles and remember not to make it too heavy as you will want to move it either regularly or at some stage. Remember to make the access as easy as possible so you can get in to clean, catch birds or collect eggs. Very occasionally secondhand housing becomes available. If you choose this option beware of disease, rotten timbers and the inability to transport the equipment in sections.

Should you already have a suitable garden shed and wish to use this, create a nesting area as in 'Nestboxes', above. Hang the feeder off the floor, about the height of the smallest bird's back. They will need a trough drinker at night. Attach a low board to the base of the doorway to avoid litter being scratched out, ideally making this removable for easy cleaning. Check that there is enough ventilation near the roof. If not, drill some 5–7 cm holes and cover them with small wire netting or remove a section of boarding and cover this with similar netting.

TYPES OF HOUSING

Information and illustrations regarding all types of housing suitable for poultry can be found in Chapter 1.

Figure 2.6 A duck house.

Top tips

▶ Choose the breed that suits the purpose.
▶ Start with just a few.
▶ Always have grass in the run or, if not enough space, use wood chippings or gravel.
▶ Provide fresh water at all times.
▶ Net over to exclude wild birds.
▶ Start without drakes.
▶ When your birds first arrive, shut them in the hut and run with food and water.

Routines

DAILY ROUTINE

Give the ducks as much time as you can as you will enjoy your
hobby more, but take just a few minutes daily to check them.

▶ *Let the ducks out after 9 am in summer and winter in order to
get all the eggs (they can be let out earlier if the run is netted
over as aerial predators are prevented).*
▶ *Change the water in the drinker.*
▶ *Put food in the trough or check the feeder has enough food for
the day.*
▶ *Collect any eggs.*
▶ *Add fresh litter if necessary.*
▶ *Put a little whole wheat in the bottom of the drinking trough
or scatter it in the covered run.*
▶ *Observe the ducks for changes in behaviour that may
indicate disease.*
▶ *Shut the pophole before dusk, checking all the ducks are in
the house.*

WEEKLY ROUTINE

This is the time to get to know your ducks and keep a closer check
on their health.

▶ *Clean the nesting area and floor and replace with
fresh litter.*
▶ *Scrape the perches if using them for Muscovies.*
▶ *Put the dirty litter in a covered compost bin.*
▶ *Wash and disinfect the drinker and feeder.*
▶ *Check that mixed grit is available.*
▶ *Ducks can go thin but their feathers cover it up. Handle any
suspicious-looking bird to check weight and condition. The
best laying ducks will be lean with their breastbone easily felt.*
▶ *Deal with any muddy patches in the run.*

Feeding and watering

TECHNICAL TERMS

Ad lib *feeding: ducks able to feed at any time (protect this from wild birds)*

Breeder pellets: *to give to the adults four to six weeks before they lay the eggs for hatching*

Chick crumbs: *high protein, small-sized feed for ducklings*

Feed bin: *vermin- and weather-proof bin such as a dustbin to keep feed in*

Feeder: *container for food to keep it clean and dry*

Free-standing drinker: *container for water to keep it clean*

Gizzard: *where food is ground up using the insoluble grit (see also Figure 1.26, page 42 for internal anatomy)*

Grower pellets: *for growing ducks*

Layer pellets: *for laying ducks*

Mixed corn: *wheat and maize combined, only useful in cold weather as very heating*

Mixed grit: *needed for the function of the gizzard*

Pellets: *a commercial ration or feed in pelleted form, grower or layer composition*

Scraps: *household food – this should only be given to ducks if it is cooked vegetable matter such as potatoes, and is good as a bribe if you need to get them to bed early*

Spiral feeder: *metal spiral at base of large bucket, pellets accessed when spiral pecked*

Wheat: *fed whole*

Ducks prefer their food liquid and will do their utmost to combine water and feed, so it is a perennial problem keeping food dry and sweet, except for wheat which can be put in a shallow trough under water.

The digestive system of a duck is very similar to that of a hen, but much faster – they tend not to store as much food in the crop as they can see at night and are still active then. Normal droppings tend to be rather liquid.

It is important that only balanced feeds from reputable sources are used. Cheap feed will be of poorer quality. Feeding scraps tends to upset the balanced ration that has been proven over many years, but green vegetable matter is appreciated in a harsh winter and to be able to call the ducks over with the reward of a small piece of fruit or stale brown bread will be very useful.

Clean water and mixed grit should be available at all times. Empty drinkers in hot weather are as bad for the ducks as frozen water in winter – they dehydrate quickly. It is important that their drinker is wide enough so that they can get their whole head in as otherwise they can get infected eyes, especially Pekins. Flint (or insoluble) grit is needed to assist the gizzard in grinding up the food, especially hard grain. From four weeks before laying commences, oyster shell or limestone grit should be provided *ad lib* to help with the formation of egg shells. Light breeds start to lay at about four months and heavier breeds at about five months. Light breed ducks will eat about 85 g (3 oz) of pellets per day plus access to wheat under water, bantam ducks need around 50 g (2 oz) plus wheat and large ducks need 120 g (4¼ oz) plus wheat, according to size.

Store feed in a vermin-proof and weather-proof bin to keep it fresh. Check the date on the bag label at purchase as freshly made feed will last only three months before the vitamin content degrades to an unacceptable level.

Refer to Chapter 1, Figures 1.19, 1.20, 1.21, 1.22, 1.23 for illustrations showing feeders and drinkers.

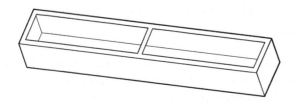

Figure 2.7 Galvanized troughs can be used for food or water.

Health, welfare and behaviour

The welfare of the birds is entirely in your hands, so if you follow the guidelines in this book you will have healthy and happy birds – they will repay your care, giving much enjoyment and fresh and delicious eggs.

Many common duck conditions or diseases can be avoided if something is understood about their behaviour. Poultry are all creatures of great habit – life is safer that way – so any change in routine can upset them. Ducks, for example, have good colour vision, so do try to wear similar colour clothes when you are looking after them and talk to them so they know you. They do recognize faces and the way you walk, but unaccustomed bright clothes will scare them.

The bird respiratory system is very different from that of a mammal – the lungs do not expand, as in mammals, but the air is pushed through them in one direction by the movement of the ribs and airsacs in a bellows motion. This is why songbirds do not appear to stop for a breath when singing and why female Call ducks can keep quacking for minutes on end.

Ducks tend not to be vaccinated against any diseases, unless there is a problem in the area they were reared. As long as they are fed correctly, stress is kept to a minimum and they are wormed about twice a year, they should remain healthy. Stress can make a duck go off its legs (the muscles seize up), so never chase them around, unless catching one with a net. Never buy a duck that has a runny nose or noisy breathing. This is caused by an organism called mycoplasma that can be carried by wild birds (see pages 203 and 216).

Insight

Waterfowl do not have much sense of taste or smell but they are sensitive to texture when digging, as they have a soft layer over their bill containing many nerve endings.

Ducks use shadows to spot potential food items on the ground and anything falling or moving is immediately investigated. All birds have colour vision and ducks are particularly attracted to red, hence the red bases to chick drinkers. Unfortunately this also means that any fresh blood is also attractive which can lead to them attacking each other, no matter how large the free-range, but once the blood has dried the danger is usually past.

The flock mentality is a protective mechanism and means that ducks must have company, even if it is only one other duck – it also means they stick together, useful if you need to herd them. A piece of wire netting or a long bamboo cane in each hand is usually enough to guide them into a new hut or different area.

There is a pecking order, not as rigid as with hens, but adding fresh stock should still be done with care and vigilance.

Anything overhead or flying is a potential predator – one bird will often alert the breeder to a sparrowhawk or buzzard and then there is a general warning buzz. Ducks comically tip one eye upwards to see something in the sky, even if it is only an aeroplane.

When greeting each other, the females dip their heads sideways and make a chuckling noise. Drakes are a serious problem in the spring as they will mate with anything that stands still. You may have seen wild mallard females with bald necks and sometimes even drowned if too many drakes are after them, particularly when they have come off their nest to feed – at that stage the males are desperate! Start without a drake and if you then decide to buy one, get only one to about six females, particularly in the light breeds. Campbell drakes, for example, are very active and will damage ducks if they have only one or two. Before you start breeding ducks, address the problem of getting rid of drakes. If they are destined for the pot, that is fine, but if you end up keeping them, the welfare of the females will be severely compromised.

Owing to their ability to see at night, ducks have to be taught to go to bed as, unlike chickens who seek shelter as the light fades, ducks are quite happy still fossicking about outside. This is, of course, dangerous where foxes are about. Fortunately ducks will be driven and keep together, but it needs to be done quietly and in daylight, otherwise they will panic if torches are used. If there is a run attached to the hut, it is merely a question of driving them into the hut, but why should they go into a dark area where they can't immediately see the reason? This is where the bribery comes in, such as brown bread or wheat.

If the hut is freestanding, make a sort of funnel to the pophole with some low wire netting, easily fixed upright with bamboo canes, and drive the ducks slowly to the hut. If one escapes, the others are likely to follow, so don't leave any gaps. Once they have the idea, all it will need is a call or handclap and off they toddle to bed.

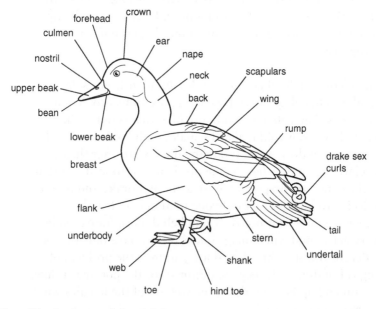

Figure 2.8 The external anatomy of a domestic duck.

TECHNICAL TERMS

Abdomen: belly
Crown: top of head
Down: the thick underlayer of feathers, traditionally used to
*   fill quilts*
Hock: 'elbow' of the leg
Moult: annual replacement of feathers
Nape: the back of the neck
Preen gland: small gland just above tail producing
*   oily substance*
Rump: lower back, immediately above the tail
Scapulars: feathers in the shoulder region, shielding
*   most of the wing when closed*
Sex curl: the curly drake's tail (not Muscovy)
Shank: leg between hock and foot
Sinus: area just below eye
Stern: area near the vent, below the tail
Vent: anus

Laying in ducks seems to be less influenced by light levels than hens as light breed ducks do lay in winter. In fact, it is a duck that holds the record for eggs in a year – 364! This was exceptional and could never be achieved by a chicken as it takes 25.5 hours to create a chicken egg whereas ducks take just 24 hours.

It is important to collect the eggs every day as the shells of waterfowl eggs are more porous than chickens' and thus bacteria can easily enter the egg. If the eggs are wanted for eating, they should be washed if dirty, with water warmer than the eggs, plus a disinfectant such as Virkon, and then stored in the fridge at a temperature of not more than 4° C (39° F). The texture of duck eggs when cooked is different from chicken eggs. The white of duck eggs seems rubbery and therefore a boiled egg is rather an acquired taste. However, using duck eggs in all other forms of cooking adds a special element of taste and moisture.

As ducks take 24 hours to create an egg, laying tends to be in the early morning, hence not letting them out until 9 am. In winter, keep the nesting area well filled with straw to help prevent the eggs getting frosted as this will crack the shell and change the protein structure, making the egg behave unpredictably in cooking and certainly unsaleable.

> ### Insight
> Ducks can moult twice yearly, especially the drakes who have drab and camouflaged plumage in the summer when they are moulting their wing feathers.

Moulting renders ducks flightless and vulnerable. Ducks spend much time preening as they need to keep their feathers waterproof. This is helped by the oil from the preen gland and also by the actual structure of the feather – keeping the barbs 'zipped up' like velcro maintains waterproofing. If feathers are missing off the back of the neck of a duck, she has been over mated, so reduce the number of drakes.

Ducks tend to nibble at everything, so be really careful if you use fencing as wire and nails will kill them by perforating the gut. Plastic string is very dangerous whether eaten or wrapped around a leg. If you see string hanging from a duck's bill, do not under any circumstances pull, but cut it off and hopefully the duck will eventually pass it.

WING CLIPPING

For information and illustration regarding wing clipping, please see Chapter 1, Figure 1.30, page 48.

PARASITES

Internal parasites
These are common in ducks that roam outside during the day. They are always on the lookout for insects and worms and some of these can contain harmful parasites. It is easy to control these

by giving a worming powder called Flubenvet to the ducks in their feed. If the ducks are in a hut and run and are moved regularly, they should be wormed two to three times a year. If they are on the same ground all the time, this interval should be shorter as they may be re-infecting themselves frequently. Flubenvet has nil withdrawal time for eggs, so it can be used at any time. It is obtainable from your vet, also see page 202. Some of the internal parasites cannot be seen with the naked eye, so worm your birds on a regular basis. Ducks have the odd habit of going lame if they have internal parasites and gizzard worm can be rapidly fatal in young birds, so be prepared. Get some Flubenvet and keep it in your medicine cupboard.

External parasites
There are several types of external parasites, all of which should be dealt with as swimming does not remove them.

▶ **Waterfowl lice:** *These live on the bird, are brown, about 4 mm (0.15") long and very thin. They hide at the base of feathers and can be seen most easily on a bird with white feathers. Control is with louse powder based on pyrethrum. Common, not life-threatening, but reduces production.*
▶ **Mites:** *The red mite is 1 mm (0.04") long, nocturnal and sucks the blood of ducks at night, while living in the hut during the day. Ducks can become anaemic, which is difficult to see, and can die. Red mite can be controlled by spraying the hut with various licensed products when the birds are outside, but the nooks and crannies, especially under a felt roof, are difficult to get at. Careful application of a blowtorch is just as effective, although time-consuming. Either treatment will probably need several applications. The red mite is grey when it is hungry and will take a meal off a human if it gets the chance. They can live for up to a year without feeding so beware second-hand duck houses!*
▶ *The northern fowl mite is a relative of the red mite but lives and breeds on the bird all the time. On a white bird, this is easy to see as a dirty mark on the base of the feathers or under the tail. Around the vent is the most common place to*

find these so, again, check here regularly. The birds become anaemic within a few days of being infested and can die. The life cycle in warm weather of both types of mites can be as short as ten days, so vigilance is really important. Treatment is pyrethrum-based louse powder or an avermectin (ask your vet) as nothing is licensed for northern fowl mite.

Both types of mite are less common on ducks than on chickens.

VICES

Ducks seem to have fewer vices compared to hens.

▶ *Over mating can be controlled by reducing the number of drakes.*
▶ *Muscovies when growing up need grass to pull at, otherwise they are likely to pull the growing wing feathers out of their friends.*
▶ *Egg eating: this will occur if an egg gets broken. Put pot eggs in the nest to discourage.*

Figure 2.9 White Campbells: not quite such good layers as the Khaki Campbells but they still have the shape of good layers.

DEATHS

It could happen that one day you find one of your birds dead in the hut. This may be because it is old or it may have had a heart attack. If a bird is found dead and you have excluded other reasons than disease as the cause of death, such as vermin, it is sensible to have it autopsied in case there is something contagious that could affect the rest of your flock. Dispose of any carcases legally and never eat a bird found dead, only ones you have killed for that purpose.

How to cope with a broody duck

Refer to the laying chart on page 75 for those breeds most likely to go broody. A broody duck has the instinct to sit upon eggs, keeping them warm and incubating them until they hatch. The duck will remain on the nest until you disturb her and then exit quacking very loudly. Ducks take less 'resetting' of the hormonal cycle than chickens, so merely moving the broody duck to another pen for a few days should do the trick. If this does not work, you may have to construct a 'sin bin' (see page 52).

Selling eggs: the regulations

Unlike the selling of hen eggs, there are currently no regulations for selling duck eggs. This could easily change, so keep an eye on the poultry press. It is important that duck eggs are not kept too long before being eaten as their shells are more porous than hens' and if bacteria has entered the shell, they will go bad quicker. If dirty, wash duck eggs in water warmer than they are with the addition of a disinfectant such as Virkon (remember, warmer water will cause the shell membrane to expand, blocking the pores in the shell). Lots of straw in the nesting area will ensure the eggs are cleaner in any case. Duck eggs should be kept in the refrigerator and used within ten days. This makes marketing them slightly less easy, but then again one of the main reasons for having home-grown eggs is the freshness factor.

Marketing of duck eggs can be to chefs, restaurants (Chinese recipes often contain duck eggs), bed and breakfasts or local farm shops. You could advertise by using a roadside sign if permissible to attract passing trade.

What to do when you want to go on holiday

For information on how to care for your ducks when you need to be away for a period of time please refer to Chapter 1, page 53.

Breeding your own stock

TECHNICAL TERMS

Batch: the number of eggs set weekly in an incubator
Breeder pellets: ration with extra vitamins, feed four weeks before expected eggs
Breeding pen: a drake and ducks of one breed, selected for a good Standard
Brooder: electrically heated area for young ducklings
Broody duck: a duck staying on the nest and incubating eggs
Chick crumbs: small-size duckling ration
Duckling: from dayold to six weeks
Embryo: developing bird in the egg
Fertile eggs: a drake must mate with the ducks
Grower: from six weeks to laying
Grower pellets: feed from four weeks to five months
Hatcher: separate thermostatically controlled insulated box where eggs are placed two days before hatch date, cleaned between hatches

Heat lamp: preferably infrared ceramic for safety
*Imprinting: instinctive reaction in first few hours of life to follow a
    moving object*
*Incubation: the time for each species to develop and hatch in the
    optimum temperature; ducks take 28 days, Muscovies 35 days*
*Incubator: electrically powered, thermostatically controlled,
    insulated box with or without automatic turning, in which
    eggs are incubated*
Layer pellets: for laying ducks
Maintenance ration: winter feeding for breeders
*Pot eggs: pottery or wooden eggs put in the nestbox to
    encourage broodiness*
*Selective breeding: only breeding from those which conform
    to the Standard*
*Set: to put eggs under a broody or in an incubator in order to
    hatch them*
Sitting: a clutch of fertile eggs

All birds naturally breed in the spring as the length of daylight
stimulates their hormones and warmer weather, and insects give their
young the best start. For the small poultry keeper there is a choice of
two methods of hatching ducks – natural or artificial. The best idea
is to gain experience and confidence in both. Remember to feed a
breeder ration to the adults for four weeks before you want to set the
eggs to increase both fertility and hatchability, following on from a
maintenance ration so that the protein is increased. Duck eggs take
28 days to hatch, with Muscovies taking 35 days. If using a Muscovy
as a broody for domestic ducks, she will not realize the change in
incubation time. Remember to wait for a fortnight before setting pure
breed eggs if the ducks have been running with another breed drake:
free-flying wild mallard will take every opportunity to mate with your
domestic birds – another good reason for netting them over.

If you do not have a drake, you will have to buy in fertile eggs
and the best contacts in the UK will be in the *British Waterfowl
Association Yearbook* and through the Domestic Waterfowl Club.
Ask if the parent stock has been fed on a breeder ration as this
will increase the number of ducks obtained. A fertile egg remains

dormant until it is placed in the correct temperature to develop. You may have the opportunity to buy in dayolds – try to make sure these have been sexed (see page 113) as it will remove the problem of too many drakes later on. It will probably cost you more to raise ducks than buying them in due to the economy of scale that rearers have, but the enjoyment will probably outweigh this.

NATURAL HATCHING

Natural hatching under a broody duck is one way to raise a few ducks and children love the experience. It is, however, essentially dependent on having a reliable broody at the same time as the eggs you want to set. It is best to leave ducks where they decide to sit, remove others from the hut and run and provide chick crumbs and water at hatching time. When the duck gets up to feed she will cover the eggs with her own down – that is how eider down is harvested in northern lands. If you try to move a broody duck as you would a broody chicken, it will probably put her off altogether. Do not let two broody ducks sit side by side as they will steal each other's eggs, try to sit on too many, letting the outside ones get cold, and probably ruin the lot. Disturbance by cat, dog or child may well upset the whole broody project so try to control access by these if possible.

If a broody duck is not available, you can use a broody chicken who will happily sit for 28 days (or even 35) without realizing it and rear the ducks (see pages 55–56). The broody is likely to get a bit upset when 'her' babies begin swimming, however. It is unlikely that you will be able to borrow a broody hen, so perhaps have a pen of Silkie crosses just for that purpose.

If your broody duck has gone broody on eggs you do not want to hatch, wait until she gets off the nest and then replace with other eggs as ducks do not really tolerate the disturbance and it would be a pity to put her off. When a broody duck gets off her nest to feed and defecate, she will do so with quite a lot of noise. This is when any males in the vicinity will jump on her, so for her own protection, keep her in a covered run or net over the area.

If you buy in dayolds, give them a drink by dipping their bills in tepid water, then place them in a smallish cardboard box so they stay warm and put them within sound of the duck or hen for about an hour. Then take them out, concealing them in your palm and place them gently under her, removing the eggs at the same time. This is best done in the dark, but it will depend on what time of day you get the ducklings. Ducks will not tolerate much interference before giving up and going off and doing something else, so go gently. The maternal instinct is strongest in the Call ducks and Muscovies; they are more settled and you will probably get away with more disturbance without upsetting them.

The duck or hen should not be disturbed for two days before the eggs are due to hatch and a chick drinker (small so they do not drown) and duck crumbs should be left within her reach. The hatch may take three days to complete, but the early ducklings need to have been able to feed. Any eggs left after three days should be gently shaken beside your ear: if they rattle they are not fertile but be careful they do not explode! Try not to disturb the broody while the hatch is on, tempting though it is to see how many have hatched, as she needs to bond with her ducklings and turn her sitting instinct into the more aggressive and protective maternal instinct. Once the ducklings have hatched, they will be very active. Try not to let them swim at this stage even though the broody duck would love to, because of the risk of getting cold and soaked. It is safer to restrict their access to water to just a drinker for the first week and then they will be strong enough to cope. Wild mallard ducklings are taken to the water by their mothers immediately they have all hatched, but domestic ducks are not quite as vigorous.

ARTIFICIAL HATCHING

For general information regarding artificial hatching please see Chapter 1, page 57.

Follow the manufacturer's instructions for an incubator and run it for a few days, checking the temperature with another thermometer. Do not add any water – it is a common misconception that in the

UK water needs adding during the incubation process. Duck eggs certainly do not need any more humidity in an incubator than hen eggs in the UK – in fact, it is possible to drown the embryos by adding too much water.

What happens if you want to set both domestic duck eggs and Muscovy eggs? They have different incubation times and it is best for them all to hatch at once, thereby keeping the humidity levels correct, so set the Muscovy eggs one week before the other duck eggs.

In order to make best use of incubator space the eggs can be candled after seven days' incubation (see page 59). Most duck eggs are easy to candle as the shells are mainly white, Cayuga and Black East Indian eggs being the exception.

Clean the incubator following the guidelines on page 60.

Take the ducklings out when they are dry and, keeping them warm and dark, transfer them to their rearing quarters and dip their beaks in tepid water. Ducks imprint on the first moving thing they see. Usually these are their siblings, so they grow up knowing they are ducks. If they imprint on humans, although cute having a duckling following you around, it is bad for their welfare as they do not know they are ducks, will not mate and will find it hard to understand why they get deserted at night.

REARING

Dayold to six to eight weeks
If a broody duck or hen is to rear the youngsters, then they will brood the ducklings and keep them warm, but you must provide overhead shelter (as they are silly in rain), chick crumbs and a suitable drinker that they cannot drown in plus food for the mother. After about four weeks, take the mother away from the growers and over a few days change their ration to the grower ration, adding a little whole wheat plus mixed grit. If ducklings stay on high protein food for too long they can develop wing joint

problems (angel wing), which makes the primaries stick out at an unattractive angle. Once this happens it is incurable, but if you notice the wing is drooping as the heavy primaries grow through, the wing can be taped (masking tape is good) in the natural position for three days which can help.

Artificially incubated ducklings need a heat lamp to keep them warm, preferably one with an infrared ceramic bulb so that they have heat and not light. The wattage will depend on the number of birds, with a 100-watt bulb being sufficient for a few and a 250-watt bulb needed for 30 ducklings. A glass bulb is likely to shatter with the way ducklings throw water about. The heat without light avoids feather pecking (Muscovies are the main culprits) as they then have natural light and darkness to maximize body and feather growth. Site the heat lamp in a draught-free place with a generous covering of shavings or newspaper on the floor or make a circle using a 2.4 m (8″) length of hardboard about 45 cm (18″) high around it. You can also use a large rectangular cardboard box and change this for each hatch. It needs to be rectangular so that the lamp is at one end and the ducklings can regulate their own temperature by moving away from the lamp. Turn the heat lamp on two days before they are due to hatch. It should be far enough off the shavings so that the temperature under it is about 39°C (102°F). If they are too hot they will scatter to the edges, panting. If the ducklings are too cold they will huddle in the middle, cheeping loudly. The ideal is to have a small empty circle just under the lamp. Transfer the ducklings from the incubator when they have dried and fluffed up. Dip their beaks in the drown-proof drinker (only use tepid water) and place them under the lamp. If they are thirsty and you give them cold water, the shock can kill them. Handling the ducklings frequently at this stage will ensure that they become tame.

Provide chick crumbs a short distance away from the lamp in a shallow container. Ducklings are unbelievably messy and will play with all the water you give them. Put the drinker on a metal grid so that the water drains through. They can stay in this area either until they outgrow it or they are weaned off the heat lamp, at about four weeks. The lamp can be gradually raised and then turned off in the middle of the day if it is hot outside, but remember to put it back on at night. The ducklings should be well feathered by this stage and able to keep themselves warm, with the lighter breeds feathering up quicker than the heavier ones.

Ducklings mature faster than chickens and the laying breeds should start to lay at about four to five months. They should be offered chick crumbs of 20–22 per cent protein, but only for about four weeks. It is probably better to get chick crumbs without a coccidiostat (a chemical to inhibit the number of coccidia (a potentially lethal parasite in the gut)) as this can upset their digestive system.

Daily routine for artificial rearing
▶ *Add fresh shavings or take out the top layer of newspaper (this may need to be done more than once a day for ducklings).*
▶ *Clean the drinker and give fresh tepid water.*
▶ *Give fresh food.*
▶ *Check the height of the lamp and the comfort of the ducklings.*
▶ *Handle the ducklings so that they get used to it and tame down.*

Rearing: six weeks onwards
The sexes will begin to be distinguished with the different voices of the ducks. As soon as they have all their body feathers they can be allowed a pond, but make sure they can get in and out of it easily. The wing feathers take the longest to grow and the males do not get their adult plumage until they are about five months old. They should be on grower pellets and some wheat at this stage, gradually increasing the proportion of wheat. Muscovies eat quite a lot of grass.

Young stock should be under observation all the time of development. Those with obvious physical defects should be removed. This will leave more room with clean houses and runs for the others to develop satisfactorily. Take precautions (by putting cardboard to round off corners) when moving stock to new houses so that they do not huddle in corners and smother. Continue to feed with best quality rations.

SEXING

Insight

Unlike hens, it is possible to sex ducks from dayold up to four weeks old.

You can sex ducks from dayold to four weeks old because they have visible reproductive parts. Hold the bird on its back with the tail facing away from you. The vent needs to be gently opened sideways by placing three fingers of each hand behind the tail (push gently upwards) and thumb and forefinger to open the vent. If the bird is male, the penis pops out, looking like a biro point. If female, inside the vent looks like an open rose. Between four weeks and four months it is very difficult to vent-sex ducks as both sexes look similar, but after four months the drake's penis looks like a spiral staircase, the female still looking like an open rose. If you do not wish to vent-sex the youngsters, you will find that the voice of male and female ducks is different – the female has a good solid quack from about eight weeks and the male sounds as though he has a sore throat, rasping. Later on, when adult plumage has come through, the curly tail of the drake is definitive, but beware unscrupulous people putting birds in a sale when they have pulled out the drake's tail!

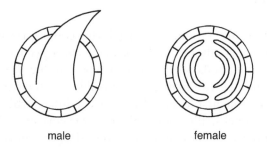

male   female

Figure 2.10 Waterfowl reproductive parts from dayold to four weeks.

BREEDING PURE

For information and guidelines on breeding ducks, please refer to Chapter 1, page 53.

Mating up
After all the ducks have passed the health and conformation tests and are considered up to Standard and fit for breeding, the question arises of how many females should run with a male. With breeding, there is no hard and fast rule about this mating ratio. The breeder's target is quality rather than quantity dayold ducklings. Thus many breeds, especially bantam ducks, are simply pair mated (one male to one female) or trio mated (one male to two females). This is very advantageous for pedigree records. In the larger breeds they will be mated in pens of six or seven females. The objective here is to get as many as possible from which to select those of high quality. Minor faults in one individual may be balanced by similar extra good points in the opposite sex. Having put the stock breeders together, eggs should be checked for shape, size, texture and colour. The better the egg, the more chance it has of producing a robust duckling, if fertile.

PEDIGREE RECORDS: THE POULTRY CLUB RINGING SCHEME

Details of the Poultry Club Ringing Scheme can be found in Chapter 1, page 67.

FERTILITY

If feeding a breeder ration, when egg laying commences under natural lighting conditions in the spring, fertile eggs may be expected within ten days of the male being introduced. If the male is already running with females it is possible that their eggs will be fertile from the first laying. If you have a different breed drake running with ducks, allow a fortnight for the correct bird to be fertile with those females after removal of the other drake. It is not necessary for the male to copulate with each female daily. He can fertilize several eggs at one time if there is free access for the sperm to travel to the ovary. The light breed females will benefit from having a rest from over-amorous drakes, too.

CULLING

Information and guidelines on culling can be found in Chapter 1, page 68.

10 THINGS TO REMEMBER FOR DUCKS

1 *Check local authority regulations.*

2 *Decide on the breed.*

3 *Get housing and pond sorted before the ducks arrive.*

4 *Get feeders, drinkers and food before the ducks arrive.*

5 *Buy good stock from a reputable source.*

6 *Check the stock is healthy.*

7 *Keep out wild birds.*

8 *Collect eggs daily, early morning.*

9 *Check the ducks twice daily.*

10 *Shut them in at night.*

3

....

Geese

In this chapter you will learn:
- *which breed to choose*
- *how best to look after your geese*
- *how to breed successfully.*

Which breed is best for you?

There are two wild ancestors of geese, the greylag (*Anser anser*)
and the swan goose (*Anser cygnoides*) and while most domestic
geese are descended from the greylag, the African and Chinese
varieties come from the swan goose.

....
Insight
Goose eggs are spectacular in cooking and infertile ones can
be used or sold for decoration.
....

Geese are kept as pets, guards, for eggs, meat and as lawn mowers.
All of the breeds will not suit all of these purposes, however,
as some of the heavier breeds lay very few eggs indeed, but all
of the breeds may be used for meat. The pure breeds are usually
available as young stock in late summer but some may need to be
booked in advance. It is still possible to obtain ordinary farmyard
geese, which are any colour and pattern but tend to be mainly
white. These may be available as dayolds or growers (best sexed).
Dayold commercial white meat geese are available in late spring.

TECHNICAL TERMS

Bean: raised, hard, oval protruberance at the tip of the upper bill
Bill: beak
Crown: top of head
Culmen: longitudinal ridge of the bill
Dewlap: fold or flap of loose skin below the lower mandible
Gander: male goose
Goose: female goose
Gosling: dayold to about four months
Gullet: throat (sometimes dewlap)
Keel: deep, pendant fold of skin suspended along the sternum
Mandibles: upper and lower bill
Neck-feather partings: all breeds descended from the greylag have these
Paunch: abdomen
Scapulars: shoulder feathers
Serrations: saw-like grooves or notches in the mandibles
Sternum: breastbone

A GUIDE TO GEESE EXPECTED LAYING CAPABILITIES

Breed	Egg colour	Numbers per annum	Maturing	Type
African	white	15	slow	heavy
American Buff	white	15	slow	heavy
Brecon Buff	white	20	quick	medium
Buff Back	white	15	quick	medium
Chinese	white	40	quick	light
Embden	white	15	slow	heavy
Pilgrim	white	20	quick	light
Pomeranian	white	20	quick	medium
Roman	white	30	quick	light
Sebastopol	white	30	quick	light
Steinbacher	white	20	quick	light
Toulouse	white	10	slow	heavy

CLASSIFICATION OF BREEDS

Heavy
African
American Buff
Embden
Skåne
Toulouse

Light
Chinese
Czech
Pilgrim
Roman
Sebastopol
Shetland
Steinbacher

Medium
Brecon Buff
Buff Back
Grey Back
Pomeranian

LIGHT BREEDS

Chinese
The Grey Chinese goose shows its descendance from the wild swan goose in colour, but the shape is very different. The Chinese, particularly the male, has a large, prominent knob above the base of the beak. This is proportionately bigger and rounder than that of the African. The neck is long and slender; there is no sign of a dewlap. Even while being a completely different species from the (greylag) domestic geese, both the African and Chinese will readily cross with the other breeds. This is common in the farmyard and likely to have contributed to the development of the German Steinbacher and several of the Russian breeds.

Although exhibited in 1845 in the UK, the recent exhibition bird has been largely developed in America and was imported in the 1970s. Light boned and short bodied, the American goose provided a small but economic bird for the table. It was extremely refined and elegant in carriage. Thus, in the exhibition pen, the modern Chinese and African geese have been cultivated to occupy extreme ends of the spectrum in size and muscularity. In addition, both are notoriously vocal and are sometimes employed in guarding property. Young

Chinese geese may lay a few eggs in the autumn of their first year, but certainly lay the most of any goose. *Colours*: the White is pure white and has an orange bill and blue eyes. The Grey has a fawn lower head, throat, neck, breast and belly, with a brown stripe from the crown down the back of the neck, back ashy brown, flanks ashy brown with paler edges, white stern, bill black, eyes dark brown. Weights: gander 4.5–5.4 kg (10–12 lb), goose 3.6–4.5 kg (8–10 lb).

Pilgrim
It is a nice story that these geese went to America aboard the *Mayflower*, but Oscar Grow from New England maintained that he named the birds 'Pilgrim' after the family's removal or 'pilgrimage' from Missouri to Iowa during the Depression. Robert Hawes has recently determined that geese were probably not aboard the *Mayflower* itself with the Pilgrim Fathers, nor on the second ship *Fortune*. The birds may have come to America by various routes: independent flocks have been found in Connecticut and Alabama, the latter with perhaps a French connection. The Pilgrim is sex-linked where the male is all white and the female completely grey apart from her stern and distinctive white spectacles. Both male and female have a dual-lobed paunch. *Colour*: the male is all white with orange bill and blue eyes, the female has white 'spectacles', the rest is light grey with back and flanks ashy grey, edged with a lighter shade, white stern, bill orange and eyes hazel. Weights: gander 6.3–8.2 kg (14–18 lb), goose 5.4–7.3 kg (12–16 lb).

Roman
According to legend, these were the Italian birds that saved the Roman Capitol from the Gauls in 365 BC. White Italian geese were imported into Britain in 1888 but these may not have been the typical Romans. They are light boned for their small size, and should carry a lot of meat. *Colour*: white, bill orange, eyes blue. Weights: gander 5.4–6.3 kg (12–14 lb), goose 4.5–5.4 kg (10–12 lb).

Sebastopol
Imported to the UK in 1859, curled feathers are the main characteristic of the Sebastopol goose, which, as its name implies, comes from the drainage basin of the River Danube.

Although predominantly a small, white bird, it is the white and buff varieties that are standardized in Britain. There are two main feather-types: the smooth-breasted and the frizzle. The first is basically a normal, smooth-feathered bird apart from exceptionally long, trailing scapular and thigh feathers that form a dense canopy of silky ribbons that often reach the floor. The second type has a much curlier appearance. The breast tends to be covered in a mass of tightly curled feathers; the wing and thigh coverts are reduced to long wisps and ribbons; even the flight feathers have lost the stiffness that would normally allow them to fly. *Colours*: the White is all white with orange bill and blue eyes. The Buff is an even buff all over, interrupted by curled feathers with orange bill and brown eyes. Weights: gander 5.4–7.3 kg (12–16 lb), goose 4.5–6.3 kg (10–14 lb).

Steinbacher

The Steinbacher is quite small but extremely attractive, with its characteristic head shape and blue-coloured plumage. The breed was recognized and standardized in 1932 in its original grey colour, then in the popular blue in 1951. *Colours*: the most popular is the Blue which is light blue-grey on head, neck, breast, back, wings and thighs with the shoulder, wing and thigh feathers edged with a lighter shade. The abdomen is silver blue. The bill is slightly convex, giving a proud look. The bean and serrations are black and there is a narrow yellow ring around the eye. Weights: gander 6.0–7.0 kg (13–15 lb), goose 5.0–6.0 kg (11–13 lb).

MEDIUM BREEDS

Brecon Buff

....
Insight
The Brecon Buff is one of the few breeds of geese originating in Britain and with its initial development well documented.
....

By 1934 the Brecon geese were breeding 100 per cent true to type and colour. The Standard was first published in 1934 in *Feathered World,* and in the 1954 *Poultry Club* edition. *Colour*: a deep shade of buff throughout with flank and scapular feathers edged with a lighter shade, white stern and dual-lobed paunch. The pink bill and feet are definitive. Eyes brown. Ganders are usually slightly paler than geese and both will fade in the summer sun. Weights: gander 7.3–9.1 kg (16–20 lb), goose 6.3–8.2 kg (14–18 lb).

Buff Back

The Buff Back is a commonly occurring colour type from the countries bordering the Baltic and the North Sea. In Britain it is similar in shape and markings to the Grey Back and, unlike the grey-back Pomeranian, it is dual-lobed in the paunch. *Colours*: the head and upper neck are solid buff, the lower neck, breast, abdomen and tail are white. The buff back is a heart-shaped, saddle-back marking. The flanks are buff with lighter edging and sometimes the buff extends under the abdomen. The bill is orange, the eyes blue. The Grey Back is grey where the Buff Back is buff. Weights: gander 8.2–10 kg (18–22 lb), goose 7.3–9.1 kg (16–20 lb).

Pomeranian

Coastal regions of the Baltic Sea, from Sweden round to Poland and Germany, are the traditional homeland of 'pied' geese. Alongside grey (wild colour) and white varieties, these geese have been recorded as early as 1550 in Pomerania (part of northern Germany and Poland). What makes the Pomeranians special, however, is their central, single-lobed paunch. *Colour*: both sexes have a dark grey head and upper neck. The lower neck, breast, abdomen and tail are white. The dark grey back is a heart-shaped, saddle-back marking. The flanks are grey with lighter edging and sometimes the grey extends under the abdomen. It is not easy getting the correct markings. The bill is orange, the eyes blue. Weights: gander 8.2–10.9 kg (18–24 lb), goose 7.3–9.1 kg (16–20 lb).

HEAVY BREEDS

African

The African is a large, imposing bird. In the most common form it is similar in colour and markings to the wild swan goose and its close relative, the Chinese. Confident, alert, loud and self-assertive, the African is inappropriately named. It is likely that its name arose from a vague and fanciful notion of its origin, but similar geese are called in China *Tse Tay*, Lion Head geese, a name that suits their proud, bulky appearance and carriage. *Colour*: the neck feathers are soft and velvet-like, and both sexes have a fawn lower head, throat, neck, breast and belly, with a brown stripe from the crown down the back of the neck. The back is ashy brown, flanks ashy brown with paler edges, white stern. Like the Chinese, they have a prominent black 'knob' above the black beak, plus there is a smooth, semicircular gullet or dewlap. Eyes are dark brown, legs orange. Weights: gander 10.0–12.7 kg (22–28 lb), goose 8.2–10.9 kg (18–24 lb).

Figure 3.1 Steinbacher family: the gander is in guard pose with his neck stretched out.

American Buff

Standardized first in America in 1947, the American Buff is a relatively large, smooth-breasted goose. Like its Brecon counterpart, it is dual-lobed. Little is known about its origins and development. *Colour*: the ground colour is diluted to a yellowish beige (buff) with a lighter shade on the flanks and scapulars, white stern. The bill and legs are bright orange, eyes dark hazel. Weights: gander 10.0–12.7 kg (22–28 lb), goose 9.1–11.8 kg (20–26 lb).

Embden

While the British may have taken to this bird in the early nineteenth century, its origins go back, at least in the area of Ostfriesland, Northern Germany, as far as the thirteenth century. How its name retains the additional 'b' dates from when the town of Emden also had that spelling in English. Like the Aylesbury, the Embdens were the focus of the Victorian breeder to produce really heavy birds for exhibition. Before 1845, they were frequently weighed in shows, dead. The heaviest was an awesome 19 kg. They are dual-lobed. Continental Embdens similarly appear more 'swan-like' than ours, with quite long heads. *Colour*: glossy white, bill orange, eyes blue. Weights: gander 12.7–15.4 kg (28–34 lb), goose 10.9–12.7 kg (24–28 lb).

Toulouse

> **Insight**
>
> The Toulouse goose was developed for table purposes in France, where it attained its celebrated reputation for *paté de foie gras* towards the beginning of the nineteenth century.

The Toulouse was imported into Britain by the Earl of Derby in the 1840s and exhibited in the first National Poultry Show of 1845. The Toulouse and the Embden were the only two goose breeds first standardized in Britain in 1865.

What marks the UK birds in particular from those elsewhere in Europe is a keel that extends down almost to the level of the underbelly and which has muscular supports that fill the usual cavities just forward of the legs. *Colour*: in both sexes, the head and neck are grey, the breast and keel a lighter grey, shading darker to the thighs. The back, wings and thighs are grey, edged with an almost white. White stern and paunch and the white tail has a band of grey across the centre. The bill is orange and the eyes dark brown. Weights: gander 11.8–13.6 kg (26–30 lb), goose 9.1–10.9 kg (20–24 lb).

Buying

Study the health points and follow the biosecurity guidelines below so that you know what to look for when going to buy birds. Reject any that do not come up to scratch: don't buy them because you feel sorry for them – they will be nothing but trouble.

BIOSECURITY FOR GEESE

- ▶ *Isolate new stock for two to three weeks.*
- ▶ *After exposure at an exhibition isolate birds for seven days.*
- ▶ *Change clothes and wash boots before and after visiting other breeders.*
- ▶ *Change clothes and wash boots before and after attending a sale.*
- ▶ *Keep fresh disinfectant at the entrance to waterfowl areas for dipping footwear.*
- ▶ *Disinfect crates before and after use, especially if lent to others. However, it is preferable not to share equipment.*
- ▶ *Disinfect vehicles that have been on waterfowl premises but avoid taking vehicles onto other premises.*
- ▶ *Wash hands before and after handling geese.*
- ▶ *Comply with any import/export regulations/guidelines.*

These are common-sense measures that can easily be incorporated into a daily routine.

POSITIVE SIGNS OF HEALTH IN GEESE

- ▶ *Dry nostrils.*
- ▶ *Bright eyes (colour varies with breed), no soreness.*
- ▶ *Clean, shiny feathers (all present).*
- ▶ *Good weight and musculature for age.*
- ▶ *Clean vent feathers with no smell.*
- ▶ *Straight toes and undamaged webs.*
- ▶ *The bird is alert and active with no sign of lameness.*

AGE OF ACQUISITION

Geese get to adult size at about five months. They are usually kept as pairs or trios due to the guarding properties of the gander. They can be sold as dayolds, growers or adult proven breeders.

> **Insight**
> It is unlikely geese will lay and breed before they are a year old, with the possible exception of the Chinese laying eggs in their first autumn but these eggs are very unlikely to be fertile.

Most breeders of exhibition stock will only sell pairs or trios, i.e. including a gander, so try to negotiate for females not quite good enough for the show pen, for instance, remembering that the heavier breeds will lay less in any case.

WHERE TO BUY

Check all stock using the health signs above before purchase.

▶ *From advertisements in smallholding magazines such as* Country Smallholding, Smallholder, Fancy Fowl, Practical Poultry, British Waterfowl Association Yearbook.
▶ *From private breeders who exhibit at waterfowl shows.*
▶ *From private breeders, ask to see the parent stock.*
▶ *At small sales. Talk to the breeder.*

WHERE TO AVOID

▶ *Large sales as prices can escalate and there may not be any history with the birds, such as age, health status and how they have been reared.*
▶ *Adverts in local newspapers may be genuine or dodgy – if it sounds good, go to see the birds before purchase.*
▶ *Car boot sales and the internet.*

Handling geese

If geese are in a house and run then driving them into the house will be the best method. Then corner the goose you want, restraining it loosely around the neck before putting your other hand over its back, confining its wings. Then slide the first hand underneath from the front, palm up, and clasp its legs between your fingers. Transfer the goose to your forearm, the other hand now on its back and its head pointing behind you with the mucky end pointing away from you – they tend to projectile defecate when picked up, so you really do not want the mess in your pocket.

Insight
> Geese can be serious about nipping as they bite and then twist, which can be excruciating, but if you have its head under your arm, as above, it can do little damage.

Beware your shins if attempting to look at a goose on her nest – the leading edge of the wing is used as a most efficient cudgel. Grasp the goose firmly around the neck at arm's length once off the nest as she will sit down and you can then pick her up as before, her head always pointing behind you. Warn any children to steer clear as a broody goose can be very aggressive.

If getting a goose out of a box or crate, loosely restrain it around the neck, put your arm over the wings, then slide your hand in under the bird from the front, palm up, and clasp the legs firmly, then transfer as above.

If geese are free-range or have a large pond, they will soon learn that being on the water is the safest place and you will be unable to catch them. Always try to be devious first and feed them away from the water so that they can then be driven into a hut or run. Once in the run, unless they have been used to being handled from dayold, it is best to catch them with a fishing landing net, then transfer them to an arm as above.

Start-up costs and other considerations

It used to be cheaper to start up with waterfowl rather than chickens, but with the threat of avian influenza (AI) and the directive to exclude wild birds from feeders and drinkers and to be able to keep all domestic poultry indoors if necessary, provision will have to be made using either a covered run or netting over an area.

START-UP COSTS

Goose house and run (e.g. for 3 geese)	£200	($300)
3 geese @ £20	£60	($90)
Feeder	£15	($22.50)
Plastic dustbin for feed storage	£5	($7.50)
Plastic movable pond	£45	($67.50)
Total	£325	($487.50)

VARIABLE COSTS PER ANNUM

Feed: 4 bags breeder pellets (for trio) for 8 weeks @ £7	£28	($42)
Wheat 250 kg (@ £4 per 25 kg)	£40	($60)
Grit, shavings or straw	£20	($30)
Total	£88	($132)

It is easy to waste feed, so get the proper feeding equipment, whether galvanized or plastic.

INCOME/BENEFITS

▶ *Maximum production from laying geese such as Chinese is 40 eggs in the spring. These can go for hatching, eating or craft work.*

- *If geese are kept in a hut then the manure can be used in your own garden or allotment or sold. It does not contain quite as much nitrogen as chicken manure, but still needs to be composted before using.*
- *Grass control in a garden or allotment (but trees need protection).*
- *Hours of observation and enjoyment!*

Housing

TECHNICAL TERMS

Apex roof: two slopes
Free-range: access to grass in daylight
Hut, house: goose house
Litter: dry and friable substrate on the floor
Nest: where eggs are laid
Pent roof: one slope
Pond: preferably movable and with a ballvalve
Run or pen: fenced exercise area, usually grassed
Shavings: livestock woodshavings for litter, also to line nest
Straw: usually wheat straw as barley straw is too soft
Ventilation: must be at roof level and above heads of birds
Window: replace any glass with wire mesh

Housing is used by geese for sleeping, laying and shelter. The welfare of the birds is entirely in your hands and certain principles should therefore be observed.

SPACE

Insight

Housing is needed for waterfowl at night for safety from predators.

The housing floor area should be a minimum of 1×1 m ($3 \times 3'$) for light breed geese, more for the heavy ones. Ideally, they will all be in a fox-proof enclosure so will not need secure housing, especially as waterfowl see well in the dark and really do not like going into huts, except to lay. Make sure the huts have fairly high interiors so the birds feel less claustrophobic if they have to be shut in at night. A hut with a large entrance door, one that perhaps drops down or at least opens sideways with no step, will encourage the birds to go in. If the area is fox-proof, then a very simple shelter for laying in is all that is needed.

WATER

Water for any geese is ideally provided in movable ponds that have a ballvalve, thus keeping the level of water constant plus avoiding the muddy patches that always seem to appear. Check the ball is screwed on tightly as geese will remove this otherwise and create a flood. A pond that can be emptied at least once a week is adequate. If a natural or dug-out pond is very large, put pea gravel around the edge to help drainage; this can be hosed down. Geese are not as destructive to green plants as ducks, but they will still have a good go at them. They find the bark of trees most attractive and can kill a tree if they are able to strip it in a circle. No matter how disciplined you think you are, waterfowl acquisition is addictive, so allow for serious expansion when planning your enclosures. Put a water trough in the hut at night if they have to be shut in. Geese do not have to swim, in fact the larger breeds tend not to, but they do like to get in the water and have a good wash.

FEEDING

Use commercial feed for the correct species and age of bird and put this in vermin-proof hoppers. If the hopper has extendable legs, such as the Parkland feeder, geese will quickly learn to use this when it is at the right height (the feeder bar should be level with their backs). Wheat can be put in shallow troughs that are filled with water so that crows and rooks are not attracted to the feed. Ideally net over the enclosure. Geese eat lots of grass, so give them a much larger area than recommended in order for some grass to remain without it all becoming a sea of mud.

VENTILATION

It is vital to provide good ventilation as the thick feathering of waterfowl is all they need for protection from cold. Housing is used to provide protection from predators rather than weather protection. Wire mesh windows or doors are best to allow good air circulation.

NESTBOXES

Geese like to choose their own nest site, but they can be encouraged by the addition of straw to an area. A simple triangular hut with no base and one side open is a good laying area for outside waterfowl as it affords some protection from aerial predators.

PERCHES

Geese sleep on the ground and do not want a perch.

POPHOLE

This is a door large enough to enable the geese to go in and out of the house at will in daylight. The most practical design has a vertical sliding cover that is closed at night to prevent fox damage. The horizontal sliding popholes quickly get bunged up with muck and dirt and are difficult to close. Light sensitive or timed gadgets are available that will close the vertical pophole if you have to be away at dusk. Most commercial goose houses have fairly high and wide popholes as geese do not like to bend to go into a hut. Geese cannot jump so make sure the access is easy.

MATERIALS

Timber should be used for the frame, which can then be clad with tongue and groove, shiplap or good quality plyboard. Recycled plastic has recently come on the market as a goose housing material with a sliding roof for good access and a good wide pophole.

It is rot-proof, light, easy to clean and less likely to harbour parasites, but check that there is enough ventilation for early summer mornings when the sun is hot and the geese are waiting to be let out.

LITTER

Wheat straw or woodshavings on the floor and for nesting areas are the best materials. Do not use hay due to the mould organisms present in it.

FLOOR

The floor can be solid, slatted or mesh. Slats should be 3.2 cm (1¼") across with a 2.5 cm (1") gap between, mesh should be 2.5 cm (1") square. If slats or mesh are used, make sure the house is not off the ground otherwise it will be too draughty. Slats or mesh make for better drainage and should be covered with straw. If the hut has a solid floor, raise the house off the ground about 20 cm (8") to deter rats but you will then need a ramp for the geese to get back in.

CLEANING

Weekly cleaning is best, replacing litter in all areas. The best disinfectant that is not toxic to the birds is Virkon. This is a DEFRA (UK Department for Environment, Food and Rural Affairs) approved disinfectant that destroys all the bacteria, viruses and fungi harmful to poultry. Remember to replace the litter in the nesting area and move all housing on a regular basis to help with hygiene.

PEN, RUN OR FREE-RANGE?

With the new regulations (see Appendix 4), provision must be made to at least net over a goose run in order to exclude wild birds. If geese are kept in a covered run, this can be moved on a regular basis so that the ground does not sour or become waterlogged. If the geese are to free-range, keep the feeder inside the house (again, to prevent wild bird access), as they will forage and eat lots of grass.

BUY OR MAKE?

If housing is bought from a reputable manufacturer and meets all the basic principles then that may be the quickest and easiest method of housing your birds. If you wish to make housing yourself, keep to the basic principles and remember not to make it too heavy as you will want to move it either regularly or at some stage. Remember to make the access as easy as possible for you to get in to clean, catch birds or collect eggs. Very occasionally secondhand housing becomes available. If you choose this option, beware of disease, rotten timbers and the inability to transport the equipment in sections.

Should you already have a suitable garden shed and wish to use this, create a nesting area with some straw. Hang the feeder off the floor, about the height of the smallest bird's back. They will need a trough drinker at night. Attach a low board to the base of the doorway to avoid litter falling out, ideally make this removable for easy cleaning. Check that there is enough ventilation near the roof. If not, drill some 5–7 cm (2–3″) holes and cover them with small mesh wire netting or remove a section of boarding and cover this with similar netting.

TYPES OF HOUSING

Information and illustrations regarding all types of housing suitable for poultry can be found in Chapter 1, page 14.

Top tips

▶ Choose the breed that suits the purpose.
▶ Start with just a few.
▶ Always have grass in the run or if not enough space, use wood chippings or gravel and then provide sprouted seeds or turf.
▶ Provide fresh water at all times.
▶ Net over to exclude wild birds.
▶ When your birds first arrive, shut them in the hut and run with food and water.

Routines

DAILY ROUTINE

Give the geese as much time as you can as you will enjoy your hobby more, but take just a few minutes daily to check them.

▶ *Let the geese out as early as you will be around. If you let them out very early in the summer, they may still be vulnerable to fox damage unless the area is fox-proof.*
▶ *Change the water in the drinker.*
▶ *Put food in the trough or check the feeder has enough food for the day.*
▶ *Collect any eggs.*
▶ *Add fresh litter if necessary.*
▶ *Put a little whole wheat in the bottom of the drinking trough or scatter it in the covered run.*
▶ *Observe the geese for changes in behaviour that may indicate disease.*
▶ *Shut the pophole before dusk, checking all the geese are in the house.*

WEEKLY ROUTINE

This is the time to get to know your geese and keep a closer check on their health.

▶ *Clean the nesting area and floor and replace with fresh litter.*
▶ *Put the dirty litter in a covered compost bin.*
▶ *Wash and disinfect the drinker and feeder.*
▶ *Check that mixed grit is available.*
▶ *Geese can lose weight but their feathers cover it up. Handle any suspicious-looking bird to check weight and condition. Broody geese lose condition quite quickly so don't let them sit for longer than 35 days.*
▶ *Deal with any muddy patches in the run.*

Feeding and watering

TECHNICAL TERMS

Automatic Parkland feeder: geese peck the feeder bar
Breeder pellets: to give to the adults four to six weeks before they lay the eggs for hatching
Chick crumbs: high protein, small-sized feed for goslings
Feed bin: vermin- and weather-proof bin such as a dustbin to keep feed in
Feeder: container for food to keep it clean and dry
Free-standing drinker: container for water to keep it clean
Gizzard: where food is ground up using the insoluble grit (see also page 42 for internal anatomy)
Grower pellets: for growing goslings
Layer pellets: for laying geese
Mixed corn: wheat and maize combined, only useful in cold weather as very heating
Mixed grit: needed for the function of the gizzard
Pellets: a commercial ration or feed in pelleted form, grower or layer composition
Scraps: household food – this should only be given to geese if it is cooked vegetable matter such as potatoes or stale brown bread, and is good as a bribe if you need to get them to bed early
Spiral feeder: metal spiral at base of large bucket, pellets accessed when spiral pecked
Wheat: fed whole

Geese are herbivores and their digestive system is very fast. Their crop is not as large proportionately as hens, so they need wheat last thing at night to see them through the long winter nights. Normal droppings tend to be rather liquid.

It is important that only balanced feeds from reputable sources are used. Cheap feed will be of poorer quality. Feeding scraps tends to upset the balanced ration that has been proven over many years, but green vegetable matter is appreciated in a harsh winter and to be able to call the geese over with the reward of a small piece of stale brown bread will be very useful.

Clean water and mixed grit should be available at all times. Empty drinkers in hot weather are as bad for the geese as frozen water in winter – they dehydrate quickly. If they do not have a pond, it is important that their drinker is wide enough so that they can get their whole head into it in order to keep their eyes clean. Flint (or insoluble) grit is needed to assist the gizzard in grinding up the food, especially hard grain.

Feed a breeder ration four to six weeks before the first eggs are expected and then chick crumbs for the goslings for the first four weeks, followed by lower protein feed plus wheat. The goodness goes out of grass around September, even though it may still be growing, so geese will need to be fed wheat plus a few pellets in the autumn and winter.

Store feed in a vermin-proof and weather-proof bin to keep it fresh. Check the date on the bag label at purchase as freshly made feed will last only three months before the vitamin content degrades to an unacceptable level.

Refer to Chapter 1, Figures 1.16, 1.19, 1.20, 1.21, 1.22, 1.23 for illustrations showing feeders and drinkers, and Chapter 2, Figures 2.4 and 2.7 showing a duck pond and a trough.

Health, welfare and behaviour

The welfare of the birds is entirely in your hands, so if you follow the guidelines in this book you will have healthy and happy birds – they will repay your care giving much enjoyment and fresh and delicious eggs.

Many common geese conditions or diseases can be avoided if something is understood about their behaviour. Poultry are all creatures of great habit – life is safer that way – so any change in routine can upset them. Geese, for example, have good colour vision, so do try to wear similar colour clothes when you are looking after them and talk to them so they know you. They do recognize faces and the way you walk, but unaccustomed bright clothes will scare them.

The bird respiratory system is very different from that of a mammal – the lungs do not expand, as in mammals, but the air is pushed through them in one direction by the movement of the ribs and airsacs in a bellows motion. This is why songbirds do not appear to stop for a breath when singing and why geese can set up an unbroken racket lasting minutes on end.

Geese tend not to be vaccinated against any diseases. As long as they are fed correctly, stress is kept to a minimum and they are wormed two to three times a year, they should remain healthy. Stress can make a goose go off its legs (the muscles seize up), so never chase them around, unless catching one with a net. Never buy a goose that has a runny nose or noisy breathing. This is caused by an organism called mycoplasma that can be carried by wild birds. It can be controlled but not cured by antibiotics. Chapter 7 contains a diseases chart, which is a summary of common poultry diseases found in small flocks.

Geese do not have much sense of taste or smell but they are sensitive to texture, especially of herbage. They use shadows to spot potential food items on the ground and anything falling or moving is immediately investigated. All geese have colour vision and goslings are particularly attracted to red, hence the red bases to chick drinkers. Unfortunately this also means that any fresh blood is also attractive, which can lead to them attacking each other, no matter how large the free-range, but once the blood has dried the danger is usually past.

Insight

The flock mentality is a protective mechanism and means that geese must have company, even if it is only one other goose – it also means they stick together, useful if you need to herd them.

A piece of wire netting or a long bamboo cane in each hand is usually enough to guide geese into a new hut or different area.

There is a pecking order, not as rigid as with hens, but adding fresh stock should still be done with care and vigilance.

Geese choose their mates in January and February, so if you need to change or add birds to the flock, do so in October and November as the pair bond is at its weakest then and new birds will be more readily accepted. Ganders, you would think, would be delighted to have more wives, but a new goose added in April can easily be bullied and driven out as the gander thinks he is protecting the rest of his flock.

Anything overhead or flying is a potential predator – one bird will often alert the breeder to a sparrowhawk or buzzard and then there is a general warning buzz. Geese comically tip one eye upwards to see something in the sky, even if it is only an aeroplane. Geese can sleep with one half of their brain at a time (dolphins do the same) so there is always someone on watch. Geese are deeply suspicious of strangers and generally will not take food from them, shouting their distrust, which also makes them good guards and burglar deterrents (something about the height of a biting beak as far as men are concerned) but they may disturb neighbours with noise.

When greeting each other, geese extend their necks and rattle their wings. They will preen each other as a bonding exercise and, once paired, there is little that will break this up. If a gander has paired with the wrong breed goose, you will have a problem as, even if separated, they will call to each other, getting very stressed. Keeping them out of sight and sound of each other may work, but they have long memories. However, if one of a pair dies, the other will readily mate up with another bird the next season.

Insight

Owing to their ability to see at night, geese have to be taught to go to bed as, unlike chickens who seek shelter as the light fades, geese are quite happy still grazing outside.

Geese staying outside at night is, of course, dangerous where foxes are about. Fortunately geese will be driven and keep together, but it needs to be done quietly and in daylight, otherwise they will panic if torches are used. If there is a run attached to the hut, it is merely a question of driving them into the hut, but why should they go into a dark area where they can't immediately see the reason? This is where the bribery comes in, such as brown bread or wheat.

If the hut is freestanding, make a sort of funnel to the pophole with some low wire netting, easily fixed upright with bamboo canes, and drive the geese slowly to the hut. If one escapes, the others are likely to follow, so don't leave any gaps. Once they have the idea, all it will need is a call or handclap and off they toddle to bed.

TECHNICAL TERMS

Abdomen: belly
Crown: top of head

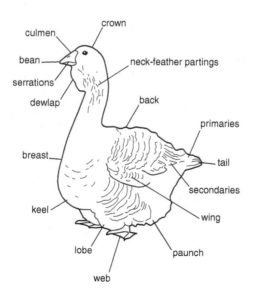

Figure 3.2 The external anatomy of a goose.

Down: the thick underlayer of waterfowl feathers, traditionally used to fill quilts
Hock: 'elbow' of the leg
Moult: annual replacement of feathers
Nape: the back of the neck
Preen gland: small gland just above tail producing oily substance
Preening: the act of feather maintenance
Rump: lower back, immediately above the tail
Scapulars: feathers in the shoulder region, shielding most of the wing when closed
Shank: leg between hock and foot
Sinus: area just below eye
Stern: area near the vent, below the tail
Vent: anus

Insight

Laying in geese begins traditionally on St Valentine's Day, 14 February, but much depends on the weather at this time.

It is important to collect the eggs every day otherwise the geese will go broody and then egg production stops. If the eggs are wanted for eating, they should be washed if dirty with water warmer than the eggs plus a disinfectant such as Virkon, and then stored in a cool place (10° C or 50° F). The texture of goose eggs when cooked is similar to duck eggs, but goose egg yolks tend to be a darker orange. Not many people would manage a whole boiled goose egg, so they are used more in cooking. Omelettes or scrambled eggs are favourites.

In frosty weather, keep the nesting area well filled with straw to help prevent the eggs getting frosted as this will crack the shell and change the protein structure, making the egg behave unpredictably in cooking, certainly unhatchable and probably unsaleable.

Geese moult once a year, usually in late summer. They spend much time preening as they need to keep their feathers waterproof.

This is helped by the oil from the preen gland and also the actual structure of the feather, keeping the barbs 'zipped up' like velcro maintains waterproofing. If feathers are missing off the back of the neck of a goose she has been over mated, so separate her by a fence from the gander for a while so she can still see and hear him. Birds like the Brecon Buff fade in the summer sun, so when they moult they regain their lovely colouring.

Geese tend to nibble at everything, so be really careful if you use fencing as wire and nails will kill them by perforating the gut. Plastic string is very dangerous whether eaten or wrapped around a leg. If you see string hanging from a goose's bill, do not under any circumstances pull, but cut it off and hopefully the goose will eventually pass it through.

WING CLIPPING

Most geese are far too heavy to fly, but occasionally a bird will take off (downhill and into the wind) when they see and hear wild geese migrating. A crash landing is usually the outcome, but if one of the smaller geese habitually gets over a fence, clip the fully grown primary feathers with scissors at the level of the small coverts on a young or adult bird. These grow back to full length the following year after the moult, usually late summer.

See Figure 1.30, page 48, for an illustration of wing clipping.

PARASITES

Internal parasites

These are common in geese that roam outside during the day. They are always on the lookout for insects and worms and some of these can contain harmful parasites. It is easy to control these by giving a worming powder called Flubenvet to the geese in their feed. They should be wormed two to three times a year. If they are on the same ground all the time, this interval should be shorter as they may be re-infecting themselves frequently. Flubenvet has nil withdrawal time for eggs, so it can be used at any time. It is obtainable from your vet, also see page 202. Some of the internal parasites cannot be seen with the naked eye, so worm your birds on a regular basis.

Geese have the odd habit of going lame if they have internal parasites and gizzard worm can be rapidly fatal in goslings when they first go onto grass, so be prepared. Get some Flubenvet and keep it in your medicine cupboard.

External parasites
There are several types of external parasite, all of which need to be dealt with as swimming does not remove them.

▶ **Waterfowl lice:** *These live on the bird, are brown, about 4mm (0.16") long and very thin. They hide at the base of feathers and can be seen most easily on a bird with white feathers. Control is with louse powder based on pyrethrum. Common, not life-threatening, but reduces production.*

▶ **Mites:** *The red mite is 1mm (0.04") long, nocturnal and sucks the blood of birds at night, living in the hut during the day. Geese can become anaemic, which is difficult to see, and can die. Red mite can be controlled by spraying the hut with various licensed products when the birds are outside, but the nooks and crannies, especially under a felt roof, are difficult to get at. Careful application of a blowtorch is just as effective, although time-consuming. Either treatment will probably need several applications. The red mite is grey when it is hungry and will take a meal off a human if it gets the chance. They can live for up to a year without feeding so beware secondhand poultry houses!*

▶ *The northern fowl mite is a relative of the red mite but lives and breeds on the bird all the time. On a white bird, this is easy to see as a dirty mark on the base of the feathers or under the tail. Around the vent is the most common place to find these so, again, check here regularly. The birds become anaemic within a few days of being infested and can die. The life cycle in warm weather of both types of mites can be as short as ten days, so vigilance is really important.*

Treatment is pyrethrum-based louse powder. The avermectins seem to be rather toxic to geese, so do not use these.

Both types of mite are less common on geese than on chickens.

VICES

Geese seem to have fewer vices than other poultry.

▶ *Over mating can be controlled by giving the goose a break from the gander.*
▶ *Goslings when growing up need grass to pull at, otherwise they are likely to pull the growing wing feathers out of their friends. Give them sods of turf if they are artificially reared indoors.*
▶ *Eggs can get broken in a nest if the goose is clumsy but geese very rarely eat eggs.*

DEATHS

It could happen that one day you find one of your geese dead in the hut. This may be because it is old or it may have had a heart attack. If a bird is found dead and you have excluded other reasons than disease as the cause of death, such as vermin, it is sensible to have it autopsied in case there is something contagious that could affect the rest of your flock. Dispose of any carcasses legally and never eat a bird found dead, only ones you have killed for that purpose.

How to cope with a broody goose

Insight
If you do not want a broody goose, remove the nest as soon as you suspect her and put corrugated tin where it was.

All geese should go broody. An alternative way to stop this is to remove the bird to another pen for a week or two. She should begin to lay again.

If broody is what you wish them to be, that is fine, but make sure that there is only one female sitting in the nesting area. If there are two, they will steal each other's eggs and probably ruin the lot. A broody goose will have laid her clutch and has the instinct to sit upon the eggs, keeping them warm and incubating them until they hatch. With the better laying breeds it is sensible to mark the first two eggs and leave those in the nest, taking each fresh egg, writing in pencil the breed and date and storing it for up to a week in cool conditions – on damp sand is good – before setting in an incubator. If the goose has gone broody while you have been away, that is the end of eggs, maybe for the whole season.

Selling eggs: the regulations

Unlike the selling of hen eggs, there are currently no regulations for selling goose eggs. This could easily change, so keep an eye on the poultry press. If dirty, wash goose eggs in water warmer than they are with the addition of a disinfectant such as Virkon (remember, warmer water will cause the shell membrane to expand, blocking the pores in the shell). Lots of straw in the nesting area will ensure the eggs are cleaner in any case.

Marketing of goose eggs can be to chefs, restaurants, bed and breakfasts, a local farm shop; alternatively use a roadside sign if permissible or put an advert in the local paper to find the craft egg decorators.

What to do when you want to go on holiday

For information and guidance on how to care for your geese when you need to be away for a period of time please refer to Chapter 1, page 53.

Breeding your own stock

TECHNICAL TERMS

Batch: the number of eggs set weekly in an incubator

Breeder pellets: ration with extra vitamins, feed four weeks before expected eggs

Breeding pen: a gander and geese of one breed, selected for a good Standard

Brooder: electrically heated area for young goslings

Broody goose: a goose staying on the nest and incubating eggs

Chick crumbs: small-size gosling ration

Embryo: developing bird in the egg

Fertile eggs: a gander must mate with the geese

Gosling: from dayold to five months

Grower pellets: feed from four weeks to five months

Hatcher: separate thermostatically controlled insulated box where eggs are placed two days before hatch date, cleaned between hatches

Heat lamp: preferably infrared ceramic for safety

Imprinting: instinctive reaction in first few hours of life to follow a moving object

Incubation: the time for each species to develop and hatch in the optimum temperature; geese take 28–32 days depending on breed

Incubator: electrically powered, thermostatically controlled, insulated box with or without automatic turning, in which eggs are incubated

Layer pellets: ration for winter maintenance

Maintenance ration: winter feeding for breeders

Pot eggs: pottery or wooden eggs put in the nestbox to encourage broodiness

Selective breeding: only breeding from those which conform to the Standard

Set: to put eggs under a broody or in an incubator in order to hatch them

Sitting: a clutch of fertile eggs

All birds naturally breed in the spring as the length of daylight stimulates their hormones and warmer weather and insects give their young the best start. For the small poultry keeper there is a choice of two methods of hatching geese – natural or artificial. The best idea is to gain experience and confidence in both. Remember to feed a breeder ration to the adults for four weeks before you want to set the eggs to increase both fertility and hatchability, following on from a maintenance ration so that the protein is increased. Goose eggs take between 28 and 32 days to hatch, depending on breed, with the heavier breeds taking the longest. If using a Muscovy as a broody for domestic geese, she will not realize the change in incubation time. Free-flying wild greylag and Canada geese will take every opportunity to mate with your domestic birds – another good reason for netting them over. It is unlikely you will be able to buy fertile eggs as they do not seem to travel well. Best contacts in the UK for young stock will be in the *British Waterfowl Association Yearbook* and through the Domestic Waterfowl Club.

NATURAL HATCHING

Natural hatching under a broody goose is one way to raise a few and children love the experience. It is, however, essentially dependent on having a reliable broody at the same time as the eggs you want to set. It is best to leave geese where they decide to sit, remove other females from the hut and run, and provide chick crumbs and water at hatching time. When the goose gets up to feed she will cover the eggs with her own down to keep them warm. Do not let two broody geese sit side by side as they will steal each other's eggs, try to sit on too many, letting the outside ones get cold, and probably ruin the lot. Disturbance by cat, dog or child may well upset the whole broody project so try to control access by these if possible, although both goose and gander are very protective of their eggs and are likely to attack an intruder of any species.

If a broody goose is not available, you can use a broody chicken (maximum five eggs) who will happily sit for 28 or so days without realizing it and rear the goslings (see pages 55–56). However, the broody is likely to get a bit upset when 'her' babies begin swimming. It is unlikely that you will be able to borrow a broody hen, so perhaps have a pen of Silkie crosses just for that purpose. Or keep some Muscovies as they will be able to cover five to seven goose eggs.

If you buy in meat geese as dayolds, give them a drink by dipping their bills in tepid water, then place them in a smallish cardboard box so they stay warm and put them within sound of the broody for about an hour. Then take them out, concealing them in your palm and place them gently under her, removing the eggs at the same time. This is best done in the dark, but it will depend on what time of day you get the goslings. Most geese (both sexes) will raise goslings whether or not they are their own and once you know your birds, there will be those, male or female, that will take on a gosling of any age and protect it.

The broody should not be disturbed for two days before the eggs are due to hatch and a chick drinker (small so they do not drown) and chick crumbs should be left within her reach. The hatch may take three days to complete, but the early goslings need to have been able to feed. Any eggs left after three days should be gently shaken beside your ear: if they rattle they are not fertile but be careful they do not explode! Try not to disturb the broody while the hatch is on, tempting though it is to see how many have hatched, as she needs to bond with her goslings and turn her sitting instinct into the more aggressive and protective maternal instinct. Warn children to stay away. Once the goslings are hatched they can have a larger drinker and if they can get in and out of a pond easily, access to a pond as their mother will keep them warm. Otherwise restrict their access to just a drinker until they are about four weeks old.

ARTIFICIAL HATCHING

Even the most experienced geese breeders can have a bad hatching season using an incubator which may or may not be affected by the weather. It is not as easy hatching goose eggs as it is other poultry, so if you have a disaster, you are in good company.

Take the goslings out when they are dry and, keeping them warm and dark, transfer them to their rearing quarters and dip their beaks in tepid water. Geese imprint on the first moving thing they see. Usually these are their siblings, so they grow up knowing they are geese. If they imprint on humans, although cute having a gosling following you around, it is bad for their welfare as they do not know they are geese, will not mate and will find it hard to understand why they get deserted at night.

REARING

Dayold to six to eight weeks
If a broody goose is to rear the youngsters, then she and the gander will brood the goslings and keep them warm, but you must provide overhead shelter (as they are silly in rain), chick crumbs and a suitable drinker that they cannot drown in plus food for the parents. Change from chick crumbs to a grower ration at four weeks to avoid angel wing, adding a little whole wheat plus mixed grit. Angel wing occurs if goslings stay on high protein food for too long. They can develop wing joint problems, which makes the primaries stick out at an unattractive angle. Once this happens it is incurable, but if you notice the wing is drooping as the heavy primaries grow through, the wing can be taped (masking tape is good) in the natural position for three days which can help. Goslings can stay with their parents until they are adult as family life is very important to them. They will need to be separated during the winter if wanted for breeding to avoid incest.

Artificially incubated goslings need a heat lamp to keep them warm, preferably one with an infrared ceramic bulb so that they have heat and not light. The wattage will depend on the number of birds with a 100-watt bulb being sufficient for a few and a 250-watt bulb needed for 15 goslings (see also Chapter 2, page 111) Handling the goslings frequently at this stage will ensure that they become nice and tame – they love putting their head under your chin and being talked to.

Goslings are unbelievably messy and will play with all the water you give them. Put the drinker on a metal grid so that the water drains through. They can stay in this area either until they outgrow it or they are weaned off the heat lamp, at about two to three weeks. The lamp can be gradually raised and then turned off in the middle of the day if it is hot outside, but remember to put it back on at night. Goslings love to go out on grass in the daytime if the weather is good, but keep them in a covered run as too much sun or sudden rain can be harmful. Without their parents to instruct them, they are a bit stupid, so you have keep a close eye on them.

Daily routine for artificial rearing

▶ *Add fresh shavings or take out the top layer of newspaper (this may need to be done more than once a day for goslings).*
▶ *Clean the drinker and give fresh tepid water.*
▶ *Give fresh food.*
▶ *Check the height of the lamp and the comfort of the goslings.*
▶ *Handle the goslings so that they get used to it and tame down.*

Rearing: six weeks onwards

The sexes will begin to be distinguished with the different voices of the goslings. As soon as they have all their body feathers they can be allowed a pond, but make sure they can get in and out of it easily. The wing feathers take the longest to grow and the males do not get their adult plumage until they are about five months old. They should be on grower pellets and some wheat at this stage, gradually increasing the proportion of wheat. See Chapter 6 for meat production.

SEXING

It is possible to sex geese in the same way as ducks from dayold up to four weeks old. This is because they have visible reproductive parts. Hold the bird on its back with the tail facing away from you. The vent needs gently opening sideways by placing three fingers of each hand behind the tail (push gently upwards) and thumb and forefinger to open the vent. If the bird is male, the penis pops out, looking like a biro point. If female, inside the vent looks like an open rose. Between four weeks and six months it is very difficult to vent-sex geese as both sexes look similar, but after six months the gander's penis begins to look like a spiral staircase, not getting to full size until the breeding season, the female still looking like an open rose. If you do not wish to vent-sex the youngsters, you will find that the voices of male and female geese are a little different – but it takes some experience and observation to get this right on a regular basis.

After a harsh winter, sometimes the phallus of a gander can prolapse and get frostbitten, which makes it wither and fall off. Always check that the males have a penis before putting them in the breeding pen.

See Chapter 2, Figure 2.10, page 113, which illustrates male and female waterfowl reproductive parts.

BREEDING PURE

For information and guidance on breeding pure geese please refer to Chapter 1, pages 53 and 65.

Mating up

The pair bond is at its weakest in autumn, so this is the time to change pairs around if you wish. It will not be successful to change partners in the spring. Check the reproductive parts of the ganders – this will probably take two people but if the bird is held belly up between the knees, head pointing behind you, one person can open the vent and check that all is well. You may think that the biting end is near a vulnerable part of your anatomy, but I have not been bitten yet using this method on thousands of geese for over 30 years.

PEDIGREE RECORDS: THE POULTRY CLUB RINGING SCHEME

Details of the Poultry Club Ringing Scheme can be found in Chapter 1, page 67.

FERTILITY

If the adults have been fed a breeder ration, fertility should be good.

CULLING

Information and guidelines on culling can be found in Chapter 1, page 68.

10 THINGS TO REMEMBER FOR GEESE

1 *Check local authority regulations.*

2 *Decide on the breed.*

3 *Get housing and pond sorted before the geese arrive.*

4 *Get feeders, drinkers and food before the geese arrive.*

5 *Buy good stock from a reputable source.*

6 *Check the stock is healthy and pair up in autumn.*

7 *Keep out wild birds.*

8 *Collect eggs as laid in spring.*

9 *Check the geese twice daily.*

10 *Shut them in at night.*

4

....

Turkeys

In this chapter you will learn:
* *which breed to choose*
* *how best to look after your turkeys*
* *how to breed them successfully.*

Which breed is best for you?

The turkey was first domesticated in Mexico by the Aztec Indians as early as 1000 AD and the first arrival in Europe of these birds was in 1524. The origin of the name is shrouded in different theories, but the Aztec name for them is *toto*. Some place names in Mexico still contain this. Turkeys were first exhibited in the UK in 1845, with a Standard appearing in the first *Book of Standards* in 1865.

....
Insight
All domestic turkeys are the same species, differing somewhat in size, but they come in a wide variety of colours.
....

Turkeys originate from America and of the five races there, the two that are the ancestors of the ones we have today are *Meleagris gallopavo gallopavo* (from Mexico, bronze with white tail border) and *Meleagris gallopavo silvestris* (from eastern seaboard of North America, bronze with brown tail border). The Mexican turkey was developed into the commercial meat type plus the paler colours and the Eastern turkey gave us the red and buff colour series.

Commercial turkeys have been selected for rate of growth and muscle mass, which gives them a waddling walk (the bronze record being 43.5 kg (95 lb))! These birds have been termed dimple or broad breasted and would be the sort of turkey seen in supermarkets. The coloured varieties, however, maintain the wild shape and activity and are known as high breasted. The meat tends to be superior in flavour in the high breasted birds as they are generally grown outside on grass. Turkey eggs can be used in cooking and all of the breeds can be used for meat production, giving varying carcase weights.

TECHNICAL TERMS

Banding: different coloured edge to feather
Barring: horizontal band of a different colour across a feather
Beak: bill
Broody turkey: hen that sits on the nest all the time to incubate eggs
*Caruncles: fleshy protruberances on head and wattles of turkeys,
    mostly stags*
Clutch: a number of eggs laid by one hen until a day is missed
Hen: adult female turkey
*Incubation: keeping eggs at the correct temperature and humidity
    so they hatch; turkeys take 28 days*
Poult: dayold to about four months
Primaries: main flight feathers
*Secondaries: the quill feathers of the wings that are visible when
    the wings are closed*
Sinus: area under eye
*Snood: extensible protruberance which hangs down in front of
    beak in stag, very small in hen*
*Spur: a projection of horny substance on the shanks of males,
    small when young and getting progressively longer with age,
    sometimes on females*
Stag: adult male turkey
Tail coverts: the feathers covering the roots of the tail feathers
*Tassel or beard: very coarse black hair growing from breast of
    older stags and hens*
Turkey Club UK: collection of turkey enthusiasts

Wattles: *the fleshy appendages at each side of the base of the beak*
Wing coverts: *the feathers covering the roots of the secondaries*

A GUIDE TO TURKEY EXPECTED LAYING CAPABILITIES

Breed	Egg colour	Numbers per annum	Maturing	Type
Bourbon Red	brown speckles on cream	50	slow	heavy
British White	cream	60	quick	light
Bronze	brown speckles on cream	70	slow	heavy
Buff	brown speckles on cream	60	quick	light
Lavender	cream	50	quick	light
Narragansett	brown speckles on cream	50	slow	heavy
Nebraskan Spotted	cream	50	slow	heavy
Norfolk Black	brown speckles on cream	70	quick	light
Pied (Cröllwitzer)	cream	80	quick	light
Slate	brown speckles on cream	50	quick	light
White	brown speckles on cream	80	quick	light

CLASSIFICATION OF BREEDS

Heavy
Bourbon Red
Bronze
Narragansett
Nebraskan Spotted

Light
British White
Buff
Norfolk Black
Pied (Cröllwitzer)
Slate
Lavender
White

Turkeys are enjoying a revival in numbers due to their personalities and choice of colours, ease of production and suitability for organic systems. A Turkey Club has been formed in the UK and has an enthusiastic following. It is the dedicated breeders who exhibit, however, as these large birds take some transporting.

Figure 4.1 Pied turkeys: mainly white with black markings.

LIGHT BREEDS

General characteristics of all colours
The head should be long, broad and carunculated with an extendable snood in the centre of the forehead, much larger in the stags. The beak is strong and curved and the eyes are bold and prominent. The throat wattle is large and pendant, but larger in the stag. The hens have sparse feathers on top of their heads and the colour of the bare skin in both sexes can range from white to red to blue, depending on mood. The body is long, deep and well-rounded with strong and large wings and a long tail containing 18 feathers. The legs are stout and strong with long claws on the toes. When displaying, the stags hold their necks towards their upright tails with the breast puffed out and make a 'poomph' noise by displacing air.

Insight

Sometimes a hen will display if agitated and it is amusing to
see little toy-like poults practising their display technique.

The stags and sometimes older hens develop a tassel that is composed
of thick, coarse, hair-like feathers growing in a bunch from the top of
the breast. This increases in length with the age of the bird.

British White

The plumage is pure white in both sexes with a black tassel. Eyes are
dark blue with a white beak and pink legs. Rarely seen at shows,
but the commercial bird is white and much heavier. The poults are pure
white. Weights: stag 9.1–9.9 kg (20–22 lb), hen 4.5–5.4 kg (10–12 lb).

Buff

Cinnamon-brown is the colour of these throughout the body
of both sexes and there is no black banding anywhere or white on
the breast of the hen. The primaries and secondaries are white
and the tail is deep cinnamon-brown edged with white. The
eyes are dark and the legs pink. The poults are pale brown.
Weights: stag 9.9–12.7 kg (22–28 lb), hen 5.4–8.1 kg (12–18 lb).

Lavender

The colour should be a pale, even shade of blue throughout
with no brown or black. Popular, but not easy to obtain.
Weights: stag 9.1–9.9 kg (20–22 lb), hen 4.5–5.4 kg (10–12 lb).

Nebraskan Spotted

The basic body colour is white with irregular red and black flecks
or spots with white wings and tail and an indistinct cream band
near the edge of it. The poults are cream and gradually acquire
the other colours as they get older. Weights: stag 8.1–11.3 kg
(18–25 lb), hen 6.3–8.1 kg (14–18 lb).

Norfolk Black

These should be a dense, glossy black throughout. There should be
no white anywhere and most specimens have slight bronze bands in
the tail, although the less of this, the better. The eyes and legs are dark

and the poults are black with a white face and some white on the underside. Weights: stag 11.3 kg (25 lb), hen 5.9–6.8 kg (13–15 lb).

Pied (Cröllwitzer)
The neck is white with narrow black banding and the body feathers are white with a black band edged with white throughout, the black tending to become broader towards the tail. The primaries are white with a black edge and the tail is white with a distinct black band followed by white at the edge. The eyes are brown and the legs horn-coloured. The poults are white, the black markings appearing as they get older. Weights: stag 9.1–9.9 kg (20–22 lb), hen 4.5–5.4 kg (10–12 lb).

Slate
Getting the slatey blue an even shade throughout is challenging as brown tends to creep in, especially in the tail. The blue can be a dark or light shade as long as it is even, but darker birds are preferred at the shows. Some strains tend to black spots in the feathers, but the less of this, the better. The eyes are dark and the legs slate. The poults are pale grey. Weights: stag 9.9–11.3 kg (22–25 lb), hen 6.3–8.1 kg (14–18 lb).

HEAVY BREEDS

Bourbon Red
The stag's head and neck are brownish-red. The rest of the body is rich, dark, brownish-red, each feather with a narrow black edging. The wings are white and the tail is white with an indistinct red bar. The hen is the same colour as the stag but with no black edging to the feathers and she has narrow white edging on the breast feathers. The eyes are brown with horn-coloured beak and legs. The poults are brownish. Weights: stag 9.9–12.7 kg (22–28 lb), hen 5.4–8.1 kg (12–18 lb).

Bronze
The body feathers are black with a broad metallic bronze band, giving the overall effect of solid metallic bronze. The hen has fine white banding on her breast. The primaries are black with distinct

white barring and the secondaries are less sharply marked. The tail feathers are black with brown barring, ending in a broad bronze band followed by a broad white band, each change of colour to be as distinct as possible. The eyes are dark, the beak horn and the legs dark. The poults are brown and dark brown striped. Weights: stag 13.6–18.1 kg (30–40 lb), hen 8.1–11.7 kg (18–26 lb).

Narragansett

This is patterned like the Bronze but the body colour is steel grey with a narrow black band on each feather, becoming solid glossy black on the back; then each feather banded with steel grey above the tail. The primaries and secondaries are distinctly barred with black and white and the tail is black barred with tan, ending in a black band edged with a broad band of pale steel grey. There should be no bronze cast in the black. Weights: stag 13.6 kg (30 lb), hen 8.1 kg (18 lb).

Buying

Study the health points and Figure 4.2 and follow the biosecurity guidelines below so that you know what to look for when going to buy birds.

Insight

Reject any turkeys that do not come up to scratch: don't buy them because you feel sorry for them – they will be nothing but trouble.

BIOSECURITY FOR FREE-RANGE TURKEYS

▶ *Isolate new stock for two to three weeks.*
▶ *After exposure at an exhibition isolate birds for seven days.*
▶ *Change clothes and wash boots before and after visiting other breeders.*
▶ *Change clothes and wash boots before and after attending a sale.*
▶ *Keep fresh disinfectant at the entrance to poultry areas for dipping footwear.*

- Disinfect crates before and after use, especially if lent to others. However, it is preferable not to share equipment.
- Disinfect vehicles that have been on poultry premises but avoid taking vehicles onto other premises.
- Wash hands before and after handling turkeys.
- Comply with any import/export regulations/guidelines.

These are common-sense measures that can easily be incorporated into a daily routine.

POSITIVE SIGNS OF HEALTH IN TURKEYS

- Dry nostrils.
- Sunken sinuses.
- Pale caruncle colour.
- Bright eyes with inquisitive expression.
- Shiny feathers (all present).
- Good weight and musculature for age.
- Clean vent feathers with no smell.
- Smooth shanks.
- Straight toes.
- The bird is alert and active.

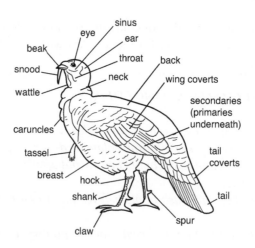

Figure 4.2 The external anatomy of a turkey.

AGE OF ACQUISITION

Dayolds and poults are generally available from May onwards. Unless you want spare stags for meat, wait to get growers until the youngsters can be sexed at about 14 weeks. Adult stock for breeding the following spring is usually available in the autumn.

WHERE TO BUY

Check all stock using the health signs above before purchase.

▶ *From advertisements in smallholding magazines such as* Country Smallholding, Smallholder, Fancy Fowl, Practical Poultry *and* Turkey Club UK Yearbook.
▶ *From private breeders who exhibit at poultry shows.*
▶ *From private breeders. Ask to see the parent stock.*
▶ *At small sales. Talk to the breeder.*

WHERE TO AVOID

▶ *Large sales as prices can escalate and there may not be any history with the birds, such as age, health status and how they have been reared.*
▶ *Adverts in local newspapers may be genuine or dodgy – if it sounds good, go to see the birds before purchase.*
▶ *Car boot sales and the internet.*

Handling turkeys

Insight
Turkeys have higher blood pressure than other poultry, so be careful when catching them.

Handling on a regular basis is very important as it is the only way to tell whether or not a bird has lost weight – even when really thin their feathers disguise this fact, so handling will give a vital early clue to any problems. Loss of weight and excess weight can be assessed by feeling the pin bones either side of the vent: they are sharp if the bird has little fat and well padded if too fat. The distance between them will indicate if the hen is laying: three vertical finger widths between the bones indicates production and less than two, the reverse.

All turkeys are immensely strong, which can make handling quite difficult, even if they have been used to being handled from dayold. Use a strong fishing landing net to catch them or drive them into a corner, but beware they do not fly onto something unreachable. Restrain the bird loosely around the neck before putting your other hand and arm over its back, confining its wings. Then slide the first hand underneath from the front, clasping its legs firmly around both shanks together. Transfer it to your forearm, with the other hand now on its back and its head pointing behind you and the mucky end pointing away from you. Unless the wings are restrained when carrying turkeys, a nasty thick lip can ensue, but they very rarely bite. Turkey muck is particularly pungent, so you really don't need it in your pocket.

If you first come across a turkey in a box, firmly grasp it around the body and wings to lift it out, then slide one hand underneath to secure the legs. Beware the strong claws.

Start-up costs and other considerations

Turkeys will use almost any type of hut suitable for chickens, as long as the perch is strong and not too far off the ground, with enough headroom and a large enough pophole for the stags.

START-UP COSTS

Henhouse and run (e.g. for 3 hens)	£200	($300)
3 turkey hens @ £20	£60	($90)
Drinker and feeder	£30	($45)
Plastic dustbin for feed storage	£5	($7.50)
Total	£295	($442.50)

VARIABLE COSTS PER ANNUM

Feed: allow 150 g of layers' pellets per bird per day,		
3 hens will eat 25 kg (one bag) in 6 weeks (9 × £6)	£54	($81)
Wheat 100 kg (@ £4 per 25 kg)	£16	($24)
Grit, shavings or straw	£20	($30)
Greens in winter	£10	($15)
Total	£100	($150)

It is easy to waste feed, so get the proper feeding equipment, whether galvanized or plastic.

INCOME/BENEFITS

► *Maximum egg production from a turkey hen is 80 eggs per year, usually in the spring and summer, but surplus can be sold for 90p ($1.35) per egg.*
► *Manure is a valuable commodity: about eight 25 kg bags of manure will be produced from three hens in a year and can either be used in your own garden or sold for about £2.50 ($3.75) per bag.*
► *Weeding of a vegetable garden or allotment.*
► *Hours of observation and enjoyment!*

Housing

Housing and space will depend to a great extent on how many birds you decide to acquire. A small garden may only support two hens whereas a larger area could have more. The basics of housing are the same no matter how large the house and commercially produced housing is usually well designed and built. See Chapter 1, page 14, for full details.

Housing is used by the turkeys for roosting, laying and shelter. The welfare of the birds is entirely in your hands and certain principles must therefore be observed.

SPACE

Floor area should be a minimum of 60 × 60 cm (2 × 2″) per bird. If you can give them more space then so much the better, bearing in mind they will be spending time in the house sheltering from the rain and wind.

VENTILATION

Correct ventilation is vital to prevent the build-up of bacteria and condensation. It should be located near the roof to ensure there are no draughts. It is often more difficult keeping the house cool than warm. In summer when it gets light early and the birds want to be out, hang up some stinging nettles in the house to keep them occupied until you get up. If they get too hot they become aggressive and can begin feather pecking each other, potentially leading to cannibalism.

Window

A mesh window would normally be located near the roof with a sliding cover to allow for adjusting the ventilation. Glass can break and does not help the ventilation. One window is best as the house can then be sited with its back to the wind.

NESTBOXES

....

Insight

Turkey hens will squeeze into impossibly small areas to lay and then have a habit of breaking the eggs, so make sure that a three-sided nestbox is at least 45 × 45 cm (18 × 18″) with straw or shavings in it, one per two hens, and located in the lowest, darkest part of the house as hens like to lay their eggs in secret places.

....

Litter in the nestboxes can be shavings or straw (not hay due to moulds). If the turkeys are laying on the floor of the hut, pin vertical 5 cm (2″) strips of black binbag (bead curtain effect) to the front of the nestbox to make it darker and more inviting. Pottery eggs can be placed in the nestbox for encouragement.

PERCHES

Perches should be broad – 5 cm (2″) square – with the top edges
rounded. They should be no higher than 60 cm (24″) as otherwise
the birds can hurt the soles of their feet when jumping down.
If you can provide a droppings board under the perches that can
be removed easily for cleaning, it will keep the floor of the house
cleaner as turkeys do two-thirds of their droppings at night.
You can also health check the droppings for colour and consistency
more easily. Normal droppings are brown with a white tip.
Every couple of hours a hen will void the contents of her caeca
(blind gut which ferments herbage) which is lighter brown
and a bit frothy.

POPHOLE

This must be large enough for the stags to enter willingly.

SECURITY

The house must provide protection from vermin such as foxes,
rats and mice. Mesh (2.5 cm/1″) over ventilation areas will keep
out all but the smallest vermin. You may also need to be able to
padlock the house against two-legged foxes, especially before
Christmas.

MATERIALS

Timber should be used for the frame, which can then be clad
with tongue and groove, shiplap or good quality plyboard.
If the timber is pressure treated by tanalizing or protimizing it
will last without rotting. The roof must be sloping to allow rain
to run off. This could be either apex (two slopes) or pent
(one slope). Avoid using felt if possible as this is where the
dreaded red mite parasite breeds. Onduline is a corrugated
bitumen that is light and warm, therefore reducing condensation.
Plywood can be used if it is treated with Cuprinol or Timbercare,
which are the least toxic wood treatments where birds are

concerned. (If you must use creosote, even the modern varieties, leave the house empty for at least three weeks to allow the toxic fumes to disperse.) To protect the plywood roof further, use corrugated clear plastic instead of felt as it lets the light through and deters the red mite which prefers dark places. Square mesh is best used on the window and ventilation areas as it is fox-proof.

Sectional construction is best for ease of moving. If building your own bear in mind that the house may need moving – a common mistake is to make it so heavy that several people are required to lift it. Wheels are available from many manufacturers to make moving their houses easy enough.

Recycled plastic has recently come on the market as a poultry housing material. It is rot-proof, light, easy to clean and less likely to harbour parasites.

LITTER

Woodshavings for livestock are the cleanest and best, straw is cheaper but check that it is fresh and clean, not mouldy or contaminated by vermin or cats. Do not use hay because it harbours harmful mould spores that will give the birds breathing problems. Litter is used on the floor, in the nestboxes and on the droppings board. Deep litter means having a friable and deep substrate such as shavings and/or straw and the turkeys scratch this over on a daily basis. It helps to break down the faeces and if properly managed has little smell. It is changed annually and makes exceedingly good compost, with the high nitrogen content of the manure breaking down the tough shavings before it is added to the vegetable garden. It should never be used fresh as it will scorch plants.

FLOOR

The floor can be solid, slatted or mesh. Slats should be 3.2 cm (1¼") wide with a 2.5 cm (1") gap between. If slats or mesh are used, make sure the house is not off the ground otherwise it will be too draughty. Slats or mesh make for easier cleaning. If the house

has a solid floor, raise it off the ground about 20 cm (8″) to deter
rats and make a dry dustbathing area. If a deep litter system is
opted for, such as using a stable or shed with an earth floor,
spread plastic under the shavings or litter to prevent damp and
previous organisms from rising.

CLEANING

Weekly cleaning is best, replacing litter in all areas. The best
disinfectant that is not toxic to the birds is Virkon. This is a
DEFRA (UK Department for Environment, Food and Rural
Affairs) approved disinfectant and destroys all the bacteria,
viruses and fungi harmful to poultry. If using the deep litter
system, this should be cleaned out once a year, but the nestboxes
should be cleaned weekly and the perches checked for a
build-up of faeces.

PEN OR RUN?

Some houses come with a detachable run or you can make your
own, preferably with a net or solid plastic roof to deter wild birds
from defecating near your turkeys and to prevent the turkeys from
flying up and roosting on the roof of the henhouse or your own
house. If movable, then the grass will stay in good condition and is
fertilized. If static, be aware how very quickly the run can become
either a sea of mud or bare of vegetation, so make it larger than
recommended or divide it into two so that each side can be rested
on a regular basis.

Some people put down wood chips (not bark as this harbours
harmful mould) to maintain free drainage. A large run should
have netting over the top to prevent wild bird access. If this is
done, the feeder and drinker can be put in the pen or run,
otherwise feeders and drinkers should be inside the house to
discourage wild bird access – not only does DEFRA consider
wild birds a risk to domestic poultry due to disease, but wild birds
will steal a huge amount of the turkey food and magpies will
quickly learn to take eggs, even from inside the henhouse.

If you are doing some gardening, then the turkey hens will love to help you find worms and insects, but they are best let out under supervision – their large eyes mean they miss nothing going on around them and they are expert at catching insects and eating weeds. Docks do not have a chance in a turkey pen, but neither do flowers as turkeys will eat almost any plant. Free-range in a domestic situation usually means daylight access to grass, not necessarily total freedom.

Beware poisonous plants such as laburnum, laurel, nightshade, but if you have children you won't have these in your garden anyway. Daffodil bulbs are toxic, so be careful of these, although most poisonous plants taste horrible to turkeys. Unless the covered run is a large area, don't attempt to plant shrubs inside it as the turkeys will soon dig these up. Clematis, honeysuckle, berberis, pyracantha or firs can be grown on the outside of the run both for shelter and to enhance the area.

If you want to weed an allotment, use a fold unit (a house and run combined) that can be moved to a fresh piece of ground as soon as the turkeys have done their job, probably daily, which means any droppings can be incorporated immediately as there will only be a few. If the turkeys are contained within the fold unit (with the feeder and drinker hanging in the run part) they will efficiently weed and manure an area of your choice and leave your precious vegetables alone, as well as being protected from the fox.

BUY OR MAKE?

This decision is the same as for chickens. Refer to Chapter 1, page 23 for full details.

TYPES OF HOUSING

There are two basic types of housing, movable and static. Movable pens are good as the birds get fresh ground regularly. Some have wheels which makes moving them easy for anyone. A disadvantage of movable pens or fold units is the limit on the size and therefore the number of birds kept in each one.

Static or free-range housing needs to be moved occasionally in order to keep the ground clean around the house – alternatively use slats or flagstones in the highest traffic area. The turkeys are allowed to roam freely or contained within a fenced-off area but you must protect their feeder and drinker from wild birds. Larger huts may have skids so they can be pulled by a vehicle to move them. Remember that tall, thin houses are unstable in windy areas, so go for something low and broad based. If a sliding or hinged roof is incorporated there is no need to have the house high enough to stand up in. When moving movable pens on a daily basis it is useful to have feeders and drinkers attached to the unit so you don't have to take the equipment out and put it all back again each time you move it.

If you have a stone or brick building that you want to use for turkeys this is obviously not movable so you may wish to go for a deep litter house, even if only in the winter. (See Chapter 1, page 28, Figure 1.16, for equipment layout.) You can let birds out or contain them in harsh weather and keeping the main door open but with a square mesh inner door will enhance ventilation and still be fox-proof. Either have hard standing outside the building that can be kept clean or place slats around the entrance to keep the feet of the birds cleaner and prevent disease-inducing patches of mud.

Commercial turkeys are often kept in a pole barn. This is a barn that has one side mesh rather than a solid wall to ensure good ventilation.

Routines

DAILY ROUTINE

Give the turkeys as much time as you can as you will enjoy your hobby more, but take just a few minutes daily to check them.

- ▶ *Let the birds out at dawn in winter and at about 7 am in summer.*
- ▶ *Change the water in the drinker.*

- Put food in the trough or check the feeder has enough food for the day.
- Collect any eggs.
- Add fresh litter if necessary.
- Scatter a little whole wheat either in the house or in the run.
- Observe the birds for changes in behaviour that may indicate disease.
- Shut the pophole before dusk, checking all the birds are in the house.

WEEKLY ROUTINE

This is the time to get to know your birds and keep a closer check on their health.

- Clean the nestbox and/or droppings board or floor.
  Put cardboard under the shavings to make the procedure easier; it will also compost down.
- Scrape the perches.
- Put the dirty litter in a covered compost bin.
- Wash and disinfect the drinker and feeder.
- Check that mixed grit is available.
- Handle any suspicious-looking birds to check their weight and condition.
- Deal with any muddy patches in the run.

Feeding and watering

TECHNICAL TERMS

Ad lib *feeding: turkeys able to feed at any time (protect this from wild birds)*
Automatic drinker: *connected to header tank or mains with valve*
Automatic Parkland feeder: *feed is accessed by pecking a coloured bar*
Feed bin: *vermin- and weather-proof bin such as a dustbin to keep feed in*

Feeder: container for food to keep it clean and dry
Free-standing drinker: container for water to keep it clean
Gizzard: where food is ground up using the insoluble grit
Mixed corn: wheat and maize combined, only useful in cold
weather as very heating
Mixed grit: needed for the function of the gizzard
Pellets: a commercial ration or feed in pelleted form, grower or
layer composition
Quill drinker: triangular shape, fed from header tank with nipple
drinkers along base edge
Scraps: household food – this should only be given to turkeys if it is
raw vegetable matter or a little stale brown bread
Spiral feeder: metal spiral at base of large bucket, pellets accessed
when spiral pecked
Turkey breeder pellets: give to the adults for four to six weeks
before they lay the eggs for hatching
Turkey chick crumbs: high protein small-sized feed for poults
Turkey grower pellets: for growing poults
Turkey layer pellets: for laying hens
Wheat: fed whole as a treat

See Chapter 1 for general feeding advice as turkeys are similar to
hens in their digestive system, but they do like to range and eat
more herbage than hens. Chapter 1 also contains illustrations of
the drinkers and feeders mentioned above, pages 34–36.

Clean water and mixed grit should be available at all times.
Empty drinkers in hot weather are as bad for the turkeys as frozen
water in winter – they dehydrate quickly. Flint (or insoluble) grit is
needed to assist the gizzard in grinding up the food, especially hard
grain. If they are not free-range, green feed is always welcomed by
the birds, but hang up vegetables and nettles to get the most benefit
from them otherwise they just get trampled.

Store feed in a vermin-proof and weather-proof bin to keep it fresh.
Check the date on the bag label at purchase as freshly made feed
will only last three months before the vitamin content degrades to
an unacceptable level.

Health, welfare and behaviour

The welfare of the birds is entirely in your hands so if you follow the guidelines in this book you will have healthy and happy birds – they will repay your care giving much enjoyment and fresh and delicious eggs plus the bonus of your very own home-grown Christmas dinner.

Many common turkeys' conditions or diseases can be avoided if something is understood about their behaviour. Poultry are creatures of great habit – life is safer that way – so any change in routine can upset them. This can range from a sudden snowfall – when they will not venture outside as the ground has changed colour – to a sudden change from meal to pellets, one of which does not look like food. This is not stupidity: their confidence is easily dented, which of course is part of the survival mechanism. The key word here is 'sudden': they will cope with most changes if they are gradually implemented.

Turkeys do not have much sense of taste or smell but they are sensitive to texture. They use shadows to spot potential food items on the ground and anything falling or moving is immediately investigated. All birds have colour vision and turkeys are particularly attracted to red, hence the red bases to chick drinkers. Unfortunately this also means that any fresh blood is also attractive, which can lead to cannibalism, no matter how large the free-range, but once the blood has dried the danger is usually past.

The birds you have bought will probably not have been vaccinated (but ask in any case) and as long as they are fed correctly, stress is kept to a minimum and they are wormed regularly, they should be healthy. Turkeys need to be wormed more often than chickens as this helps to remove the vector for a fatal disease of turkeys, Blackhead, which is carried by intestinal worms that chickens commonly have.

Swollen sinuses, a runny nose or noisy breathing are caused by an organism called mycoplasma that can be carried by wild birds. It can be controlled but not cured by antibiotics. Chapter 7 contains a diseases chart. It is a summary of common poultry diseases found in small flocks. It is essential to involve your veterinary surgeon (take this book with you) if you have problems with your poultry and although some wormers and louse powder can be obtained through licensed outlets (such as agricultural merchants), most drugs and medicines are only obtainable through a vet. Wash your hands after handling medicines and observe the withdrawal instructions on the labels of drugs, so do not eat eggs or birds when medicines are being given. If medicines are given in water, make no other water available. Most diseases are management related, for instance rats and mice carry some diseases as well as all those carried by wild birds, therefore many diseases can be prevented by good management.

See Chapter 1 for internal anatomy and reproduction as turkeys are very similar to chickens in this respect, except for the capacity and efficiency of the semen storage glands. The record of production of a fertile egg after a male has been removed from the female belongs to a turkey and is 72 days!

Turkeys are insatiably curious and will get into all sorts of mischief: they have strong hooligan tendencies. Their boredom threshold is low and if there is nothing else to do, they will pick on one of their own number and have a mugging session.

The alarm call is a 'putt, putt', given if a strange noise is heard or a stranger spotted. Small children love the noise competition!

When fighting each other, turkeys lock on with their beaks and use their strong legs and claws. If a stag decides that he dislikes a person or just people generally, he can be dangerous as he kicks out and clouts with his wings. A solution involving the cooking pot is really the only answer as turkeys have only one idea in their heads at a time and it is difficult to change this. If transporting them any distance, be aware that their droppings are particularly pungent. They live for about ten years and should lay for about six of these.

Plastic string is very dangerous. If you see string hanging from a turkey's beak, do not under any circumstances pull, but cut it off and hopefully the bird will eventually pass it. String (such as that closing feed bags) can get wrapped around the leg or foot of a bird causing swelling and ultimately gangrene, so get into the habit of picking it up.

Turkey hens fly well, particularly when looking for a nesting site, so a clipped wing (Figure 1.30, page 48) is useful here. Young stags can fly a bit, but they get more sedentary as they get heavier. However, a 1 m (3″) high fence is easily jumped/flapped onto by stags of any age. All turkeys love to perch at night when roosting and if not given a perch they will get up onto inappropriate heights, such as roofs. Their flock instinct is powerful, which makes them good guards in daylight, but they do not see well in the dark, unlike waterfowl. This flocking makes them easy to move by driving, if done slowly.

Turkeys moult once a year, usually in late summer, but the backs of the hens can become bare of feathers due to the feet of the stag. It is usual to have a turkey stag with two or three hens in the breeding pen. Stags are rather slow at mating and spend a long time treading (literally) on the backs of the hens. This can lead not only to feather damage but can split the flanks of the hens severely. It is best to fit the hens with a leather or strong cloth 'saddle' for the breeding season. These can be obtained from various sources advertising in

Figure 4.3 A turkey breeding saddle; the correct position on a turkey hen.

the poultry press. The saddle is shaped like an apple with a loop of material on each 'shoulder' that goes around the wings, with the bulk of the material protecting the back and flanks of the hen.

PARASITES

Internal parasites
These are common in turkeys that roam outside during the day. They are always on the lookout for insects and worms and some of these can contain harmful parasites. It is easy to control these by giving a worming powder called Flubenvet to the turkeys in their feed. Flubenvet has nil withdrawal time for eggs, so it can be used at any time. With the risk of the debilitating disease Blackhead, it is sensible to worm turkeys regularly to reduce this. Flubenvet is obtainable from your vet, also see pages 201–212.

External parasites
There are several types of external parasites, all of which need to be dealt with as dustbathing will only remove a few.

▶ **Lice:** *These live on the bird, are yellow, about 3 mm (0.12") long and lay eggs (nits) at the base of feathers, usually under the tail, so this is a place to look on a regular basis. Control is with louse powder based on pyrethrum. Not life-threatening, but reduces production.*

- **Mites:** *The red mite is 1 mm (0.04") long, nocturnal and sucks the blood of turkeys at night, while living in the hut during the day. Turkeys become anaemic with a pale face and can die. Red mite can be controlled by spraying the hut with various licensed products when the birds are outside, but the nooks and crannies, especially under a felt roof, are difficult to get at. Careful application of a blowtorch is just as effective, although time-consuming. Either treatment will probably need several applications. The red mite is grey when it is hungry and will take a meal off a human if it gets the chance. They can live for up to a year without feeding so beware secondhand henhouses!*
- *The northern fowl mite is a relative of the red mite but lives and breeds on the bird all the time. On a white bird, this is easy to see as a dirty mark on the base of the feathers or under the tail. Around the vent is the most common place to find these so, again, check here regularly. The birds become anaemic within a few days of being infested and can die. The life cycle in warm weather of both types of mites can be as short as ten days, so vigilance is really important. Treatment is pyrethrum-based louse powder or an avermectin (ask your vet) as nothing is licensed for northern fowl mite.*
- *Scaly leg mite burrows under the scales of the legs making raised encrustations and is very irritating for the bird. Treatment is by dunking the affected legs in surgical spirit once a week for three weeks.*

VICES

Sadly, turkeys can acquire vices but these are often due to management and husbandry problems:

- *Egg eating happens if an egg gets broken in the nestbox. This could happen if there is not enough litter, the nest is too small or the shell of the egg is weakened for some reason or even missing. Turkeys know that eggs are extremely good food and will take any opportunity to consume an egg once a shell is broken. Prevention is best, such as covering the front of the nestbox with vertical binbag strips, providing enough litter and enough grit, putting several ping-pong balls, golf balls or pot eggs in the nestbox and on the floor of the hut. Find out why the eggs have been breaking.*

▶ *Feather pecking can begin if the growing birds get too hot or if there are too many in a brooder. Reduce the temperature. You may also have to take the worst offenders to another brooder. In adult birds, feather pecking is often caused by boredom so hang up some nettles to give entertainment. If tails get pecked often enough then the feather follicle is damaged and will probably never grow back properly.*

▶ *Roosting in the nestbox and therefore defecating in it: cover over the nestbox entrance during the afternoon in the hope that the offender will decide to roost instead. Make sure there are enough perches for all the birds to sit in comfort as one may have got bullied off the perches.*

DEATHS

It could happen that one day you find one of your turkeys dead in the hut. This may be because it is old or it may have had a heart attack (if the comb is purple on a hen or the head is dark on a stag). If a bird is found dead and you have excluded other reasons than disease as the cause of death, such as vermin, it is sensible to have it autopsied in case there is something contagious that could affect the rest of your flock. Dispose of any carcasses legally and never eat a bird found dead, only ones you have killed for that purpose.

How to cope with a broody turkey

Prevention is the most effective method by collecting eggs every day. A broody hen turkey has the instinct to sit upon eggs, keeping them warm and incubating them until they hatch. Any of the colours will do this, given half a chance. You will find that one hen will insist on staying most of the time in the nestbox and may well try to peck you when you are collecting eggs. To check if she is really broody, gently slide your hand under her, palm up, and if she 'cuddles' your hand with her wings, then she is serious. The best way to reset her cycle is to construct a 'sin bin', see Chapter 1, page 52. After two weeks in the sin bin, integrate her carefully back into the flock (or place the sin bin in the henhouse if there is space) and she should begin to lay properly again after that quite quickly, therefore earning her keep. The next time she looks as though she is going broody, put her in the sin bin straight away for a few days. This may be enough as the hormones are not at full power at the beginning of the broody cycle.

Selling eggs: the regulations

Unlike the selling of hen eggs, there are currently no regulations for selling turkey eggs. This could easily change, so keep an eye on the poultry press. If dirty, wash the eggs in water warmer than they are with the addition of a disinfectant such as Virkon (warmer water will cause the shell membrane to expand, blocking the pores in the shell). Lots of straw in the nesting area will ensure the eggs are cleaner in any case.

What to do when you want to go on holiday

For information on how to care for your turkeys when you need to be away for a period of time please refer to Chapter 1, page 53.

Breeding your own stock

Insight
The eggs from turkeys are larger than chicken eggs, but shaped similarly with a broad end and a distinct pointed end. The shells are cream coloured and have varying amounts of brown speckling and spotting on them.

See Chapter 1 for general breeding information, as turkeys are hatched and reared in a similar way to chickens and can be incubated. As in chickens, the eggs increase in size as the birds get older and they are good to eat boiled, having a texture the same as a chicken egg, and being larger, go further in cooking. Hens will lay between 50 and 90 eggs a year, mostly between March and July.

The eggs take 28 days to hatch and turkey hens make good mothers if allowed to sit, but do not let two turkeys sit side by side as they will steal each other's eggs and probably ruin both clutches. Most broody turkeys are best left where they decide to sit as they will go off being broody if they are moved. Remember to move the other occupants out of the hut and run as the stag may decide to mate with the broody, thus disturbing her, and the other hens will want to lay their eggs where she is sitting. If you buy in dayolds, give them a drink by dipping their beaks in tepid water, then place them in a smallish cardboard box so they stay warm and put them within sound of the hen for about an hour. Then take them out, concealing them in your palm and place them gently under her, removing the other infertile eggs at the same time. This is best done in the dark, but it will depend on what time of day you get the chicks. Remember you have only 24 hours if you wish to add more: she cannot count but she has colour vision and can tell the difference between poults.

ARTIFICIAL REARING

Insight

Dayold turkeys are very cuddly and love human company.

Dayolds are not difficult to rear as long as proper turkey food is used as this contains the correct level of protein. They love playing games and sparring as they get older, and they need shade in hot weather. Sexing is not easy until the stags begin to grow away from the hens, becoming larger at about 12–14 weeks. The hens retain feathers on their heads for longer and the earliest they can lay is about 30 weeks, but normally it is the following spring after hatching. See Chapter 6 for meat production.

Daily routine for artificial rearing

▶ *Add fresh shavings or take out the top layer of newspaper (if this needs to be done more than once a day the area is not large enough for the number of poults).*
▶ *Clean the drinker and give fresh tepid water.*
▶ *Give fresh food.*
▶ *Check the height of the lamp and the comfort of the poults.*
▶ *Handle the poults so that they get used to it and tame down.*

SELECTION FOR BREEDING PURE

Chapter 1, page 65 contains full information relating to selection.

CULLING

Information and guidance on culling can be found in Chapter 1, page 68.

10 THINGS TO REMEMBER FOR TURKEYS

1 *Check local authority regulations.*

2 *Decide on the breed.*

3 *Get housing sorted before the turkeys arrive.*

4 *Get feeders, drinkers and food before the turkeys arrive.*

5 *Buy good stock from a reputable source.*

6 *Check the stock is healthy.*

7 *Keep out wild birds.*

8 *Collect eggs as laid in spring.*

9 *Check the turkeys twice daily.*

10 *Shut them in at night.*

5

Other breeds – guinea fowl and quail

In this chapter you will learn:
- *if guinea fowl or quail are for you*
- *how best to look after your birds.*

Guinea fowl

There is just one type of domesticated guinea fowl (*Numidia meleagris*) available in the UK and elsewhere and these are used mainly for meat (see Chapter 6).

> **Insight**
> Egg production is not particularly good at around 50 per year and dayold keets (youngsters) are only available at certain times.

Guinea fowl can be treated as chickens in the rearing and feeding but they are able to fly once nearly adult, so it is essential to wing clip or net them over. They do not take kindly to being kept indoors and will roost as high as possible, shouting at anything strange. They will get some of their food by foraging, but do like some layer pellets and wheat.

Sexing guinea fowl is by call only: the female shouts 'get back, get back' with two syllables and the male has a single-syllable screech. The naked heads and pendant wattles can vary in size but this is not always linked to the sex of them. The colours are pearl (dark with small white spots), lavender (pale blue with small white spots) and then some others have varying areas of white.

Figure 5.1 Guinea fowl: attractively spotted but rather noisy.

Insight

Occasionally adults are available but what must be taken into account is that these are very, very noisy birds, so neighbours have to be considered.

Coming originally from Central Africa, the shells of the eggs are very tough to avoid water loss and hatching can sometimes be a problem. Incubation time is 28 days, but due to the strong shell, storage of two to three weeks does not seem to affect hatchability unduly.

I have had guinea fowl hatch and rear their own but they need to be contained once hatched as the mother will walk off into long wet grass and the keets will chill from the wet grass and die just trying to keep up. If contained, they make quite good mothers, albeit a bit scatty, so it is probably best to artificially incubate and rear them. See Chapter 1 for detailed information regarding this.

In a flock, they remind me of a group of busybody spinsters, shouting their disapproval at the youth of today as they rush about with their skirts trailing, getting into everyone's business – entertaining but noisy.

Other types of guinea fowl such as helmeted or vulturine are not domesticated although they are sometimes kept in captivity.

Quail

It is the tiny Japanese quail (*Coturnix coturnix*) that is used for egg production.

> **Insight**
> Quail can be sexed by colour if they are the fawn variety, the males having a plain breast and the females spotted feathers.

Another colour apart from fawn is Tuxedo quail, which are white and dark brown with little difference between the sexes except that the male usually has foam at his vent and the female has a larger abdomen for egg production.

Quail are not kept outdoors as they are very susceptible to damp conditions and get pneumonia easily. They are kept either on shavings or on wire in cages. They scratch a lot, so if on shavings, the drinker needs to be either the nipple type or put on a grid to prevent clogging with shavings. The cages have integral feeders and drinkers.

Quail mature incredibly quickly – at six weeks of age the first eggs arrive. Providing hardboiled quail eggs for a local pub or restaurant is an option as quail will lay almost continuously for about a year. (Peel them by steeping the cooked eggs in vinegar, which dissolves the shell). The adults, once at the end of laying, can be used for the pot as they are still tender at this stage. Obviously replacements will have to be acquired or bred.

Figure 5.2 Japanese quail: the male is calling (his head is highest) and his breast is plain unlike the females who have spotty breast markings.

Insight

Quail are unreliable as mothers and in fact drop their eggs anywhere, not even making a nest, so artificial hatching and rearing is the norm.

Incubation is 17 days and small incubators do the job well. The chicks are incredibly small and active and tend to leap out of the incubator once it is opened. They also like to try to drown, so put pebbles in the base of chick drinkers to avoid this and make absolutely sure they cannot jump or climb out of the brooder area. You may have to reduce the size of ordinary chick crumbs for the first week or two by putting them through a blender. Adult quail live on chick crumbs and millet.

Handle quail by either catching them in a small net or driving them into a corner. They fly well, so close off any escape routes. Hold them around the body to contain the wings.

Other types of quail (44 species) are not domesticated although are sometimes kept in captivity.

WHERE TO BUY AND WHERE TO AVOID

The principles are the same as for chickens. See Chapter 1, page 10.

10 THINGS TO REMEMBER FOR GUINEA FOWL AND QUAIL

1 *Check local authority regulations, remember guinea fowl are noisy.*

2 *Decide on which species.*

3 *Get housing sorted before the birds arrive.*

4 *Get feeders, drinkers and food before the birds arrive.*

5 *Buy good stock from a reputable source.*

6 *Check the stock is healthy.*

7 *Keep out wild birds.*

8 *Collect eggs as laid.*

9 *Check the birds twice daily.*

10 *Shut them in at night.*

6

Meat production

In this chapter you will learn:
- *how to grow meat birds*
- *how to slaughter humanely*
- *how to prepare a bird for the table.*

General principles

> **Insight**
>
> Any bird destined for the table should not be given a name –
> you do not eat your friends – unless they are called paxo, sage
> and onion or something equally as evocative.

> **Insight**
>
> There are many meat hygiene regulations if you try to sell a
> processed bird.

These instructions are to enable you to produce food for your
own family. The hygiene regulations are certainly achievable,
but I strongly suggest you get experience on a small scale first.

Three to four weeks before slaughter, birds of any species
should have been fed on a ration that does not contain any
medication, known as a withdrawal ration. A fattening ration
could contain medication, so always check the label on the
feed bag.

Insight

Birds for slaughter should be starved for 12 hours but still allowed access to water.

Starving for 12 hours (with access to water) ensures that the gut is nearly empty and therefore that poultry feed cannot spoil in it. The birds need to be handled with great care as any bruising will be obvious after processing.

Processing birds for the table is quite hard physical work, so plan your time – it is easier if only a few are done at once, so just slaughter those you know you can pluck and clean in the appropriate time. If you have not done this before, do one bird at a time. It does get easier with practice.

CHICKENS

If you have bred some chicks there are inevitably spare cockerels. Separate these from the females and run them together in a fairly restricted area so they can fatten up a bit.

Insight

If they bicker, add an older cock bird to keep the peace, or provide extra entertainment such as hung up nettles.

Pure breeds can take six to ten months to get to an edible size, so allow for this time by a suitable hanging period to tenderize the meat – 2–3 days in warm weather, up to 7 in the winter. If broiler chicks are available in your area (see *Taking it further*, Useful addresses) these can be reared on grass and are edible from eight weeks of age, needing very little hanging time. They can be slaughtered over a period of time as the males and females grow at different rates. Do not be tempted to keep some back to breed your own as these will just continue to grow and become unable to walk. These broilers are the end product, not the breeding lines.

Indian Game cross Dorking has been featured on the TV programme, River Cottage, and although good tasting, it takes about ten months to get there. Any of your spare cockerels of any breed would be as tasty, due to the life they have led, but the distribution of the meat goes more towards the legs, unlike broilers.

WATERFOWL

Spare drakes can be kept together in a fairly restricted area and fed on barley or wheat to fatten them a little. They are edible from about five months old, particularly the heavier types. Check that they are through a moult before slaughter as the pin feathers are a nightmare to remove. There is a short to no hanging period.

Geese are sometimes reared on grass and fattened on barley for Michaelmas, but Christmas is a good time as well. Check for pin feathers as above and hang for five to seven days.

> **Insight**
> The fat from a goose is spectacular for roasting potatoes in throughout the following year.

TURKEYS

If young stags are to be kept together for meat production, it is sometimes useful to put an older stag in with them to keep the peace. Turkeys can be killed at any time from 22 weeks old for meat, as long as there is a bloom of fat under the skin and the feathers are through a moult. If breeding for Christmas, plan carefully and hang for at least a week.

GUINEA FOWL

These are produced commercially for meat, so mature quickly. They are edible from 12 weeks old and will need very little hanging. They have a slightly gamey taste.

QUAIL

These are ready for the table when they have finished laying, at about 14 months old. Very little or no hanging is needed.

Slaughter

This needs to be done out of sight and sound of other poultry. On a small scale, the humane way to kill a chicken is by dislocating the neck. There are two sets of blood vessels in the neck and only by dislocating it can you disrupt blood and nerve supply to the brain and therefore first render the bird unconscious and then shortly afterwards, dead. Neck dislocation should be carried out only if immediate unconsciousness is induced without causing pain or suffering. Small numbers of birds on home premises can be killed by neck dislocation without prior stunning – this could take two people with a large goose or turkey. (Stunning is usually done in slaughterhouses with one electrode on the overhead line where the birds are hung and the other in a tank of water where the birds' heads are dunked, thereby rendering them unconscious before they are bled.)

There is also a metal device for large birds that is kept screwed to a wall – a lever is pulled down to dislocate the neck, which is held on a platform. Killing cones are also commercially available, which

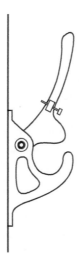

Figure 6.1 A humane killer for larger birds.

avoid bruising of the wings and carcase for turkeys – the birds are placed into the cone head down and then the neck is dislocated.

Airguns can be used at close range on large birds, by putting the barrel at the back of the skull with the bird's head on the ground, then dislocating the neck to allow space for the blood to pool. Cutting the throat or removing the head is horribly messy. Remember – it is the responsibility of the keeper to ensure that poultry are killed humanely.

Plucking and hanging

Current advice is to bleed the bird by hanging it by its legs and letting the blood pool in the neck cavity, created by the dislocation. (Funny how game pheasants are not bled but are suspended by the neck and still taste very good.)

Insight
Once dead, but while still warm, pluck the body feathers all over and to half way up the neck of the bird. Take just two or three feathers at a time and pull sharply against the direction of growth.

A warm carcase means that the feathers not only come out quite easily, the skin is stronger and tears less easily. Waterfowl have two layers of feathers and Muscovy ducks have three layers, so dip these in scalding water for 60 seconds and then the feathers come out easily. You will need quite a large receptacle to do this for geese. Alternatively, waterfowl can be ironed (with an old iron) over an old tea towel as the heat, in the same way as the scalding water, loosens the feathers.

Another suggestion is to dip the bird in hot paraffin wax (a certain amount of specialized equipment is needed here) which, when cool, is peeled off with all the feathers. The wax can be remelted and used again, once the feathers have been strained off. Plucking is a personal preference and will also depend on how much time you have – a chicken would take about 20 minutes plucked by hand by an experienced person.

Most people recommend plucking the whole wings, but there is not much meat on them and it takes too long – on a large bird I might pluck the secondaries (pliers are useful here) and leave the rest to be cut off when the bird is gutted.

If you have tried hand plucking and consider it prison work, there are electric plucking machines on the market, which, although expensive, are a joy to use, with a chicken taking about two minutes. However, most of these will not successfully remove the body feathers of a goose or wing/tail feathers of any poultry.

Shot game pheasants are hung for about a week in winter. If you wish to hang birds, do so after they have been plucked but not gutted. The timing is, again, personal preference, but in summer I would hang a six-month chicken in a cool fly-proof place for two days; in winter, a week. This helps to break down the muscle fibres and tenderize the meat.

Insight

Older birds may need a longer hanging period. Supermarket broiler chickens are not hung at all, but as they are only 39 days old this does not matter. On a small scale, birds are likely to be very much older than this, therefore hanging can be useful.

To make a simple fly-proof larder: take a melamine bookcase, about 60 x 60 x 20 cm (24" x 24" x 8") deep, with the middle shelf removed. Cut a broom handle to fit and fix near the top. Make short wire s-shaped hooks to hang over this and then hang your poultry with string from these hooks. For the open front of the bookshelves, get a fly screen window kit that is fixed with stuck velcro so that when in place, strips of velcro hold the mesh to the melamine edges and no flies can get in. Put in a cool place.

> How much meat can you expect to get? The general rule
> is that an oven-ready bird weighs 70–75 per cent of its live
> weight, but of course the bones are part of this.

If you wish to consume older birds that would really only be
suitable for casseroles, skin them instead of plucking. After
hanging, make a shallow cut over the breast and peel back the skin
up to the neck and down to the hocks. With a sharp knife, take off
the breast meat by cutting along the breastbone then, in a slight
arc, down towards the wing and along the wishbone. If you can
be bothered to remove the tendons from the legs, this will make
more tender meat, but if not, cut the leg off at the joint. Then press
the leg away from the body until you can see the hip joint, and cut
the leg meat where it joins the body. This is a speedy method and
can be used on any of the poultry species, but the cooking should
involve some liquid otherwise the meat will dry out. Either an Aga
or a slow cooker is the best method used for older birds.

Processing or cleaning

EQUIPMENT

The following equipment is required if you intend to clean your
own poultry:

▶ *rubber gloves*
▶ *disposable cloth*
▶ *sharp knife*
▶ *kitchen scissors*
▶ *blowtorch or methylated spirits*
▶ *string (not plastic) for trussing.*

When the required hanging time has elapsed, prepare to gut the
bird in your chosen, clean, processing area. Stainless steel is a
good surface on which to do this as it can be cleaned with bleach
both before and after the processing. Melamine is second best.

Plastic surfaces can score from the knife and hold bacteria and wood (unless a butcher's block) is uncleanable.

The remaining feather stubs or fine hairs need to be removed, either with a blowtorch or with lit methylated spirits in a small metal container. Either way, this singeing will smell pretty awful.

Cut off the part of the wing that still has feathers by slicing down through the flap of skin to the elbow joint with the knife and then bend the joint open and cut or snip away.

The legs are next. Break the centre of the shank on a sharp edge (such as a table). Twist the shank and pull it at an angle so that the bone sticks out making a T-shape with the rest of the leg. Pull this so that the tendons come out from the thigh area. This is good for your pectoral muscles! For large birds, there is commercial gadget called a Sinupul (Bingham Appliances) that I would not be without as with neat leverage the tendons are removed from the largest turkey or goose.

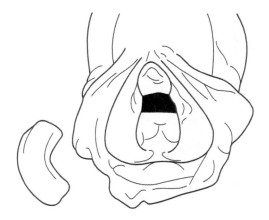

Figure 6.2 Gutting – bird breast down, cut back skin of neck to shoulders, neck giblet and crop removed.

Cut off the head and upper neck with the scissors and turn the bird onto its breast. Slice through the skin to between the shoulders with the knife, exposing the neck giblet. Work this free of the underlying crop and cut off at the shoulders, keep for gravy. Work around the

trachea and oesophagus with your fingers, then cut these close to the
shoulders. Pull the crop and any other tissue away from the breast
skin and discard. The bird can be stuffed in this cavity later if wanted.

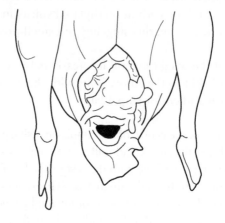

Figure 6.3 Gutting – teardrop shape cut around the vent.

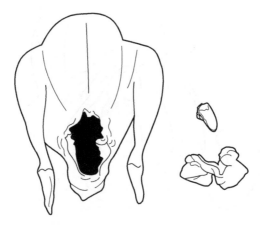

Figure 6.4 Gutting – guts removed, heart and liver kept.

With the sharp knife, cut around the vent in a teardrop shape.
Work your fingers into the body cavity and pull out all the
intestines including the heart. Cut through the muscle of the
gizzard and peel off the inner lining, keeping the purple muscular

part for gravy. Keep the heart and liver as offal but make sure that the dark green gall bladder is carefully removed intact so that it does not taint the liver. Discard the intestines. If necessary, wipe the cavity with clean kitchen paper before stuffing.

Trussing

With the bird on its back and its tail towards you, take a metre (yard) length of string and tie a single overhand around the parson's nose at the centre point of the string. Take each long end over the top of the hock joint and back towards the parson's nose, pull tight, then take the string up and over between the leg and the body, flip the body over and tie at the back, trapping the folded-over breast skin. Then take the string to the front and tie on top of the breast.

Bag, remove air, tie, label and freeze. Thaw slowly (overnight in the fridge) before cooking.

Insight

This should be good up to a year in a domestic freezer.

Or eat freshly roasted. It still gives me a kick to have produced an entire meal with homegrown meat and homegrown vegetables.

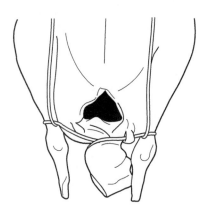

Figure 6.5 Trussing – string secured around parson's nose, over hocks and along inside leg.

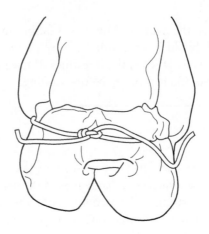

Figure 6.6 Trussing – string tied at back of bird to secure neck skin.

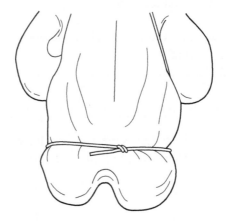

Figure 6.7 Trussing – final knot over breast.

10 THINGS TO REMEMBER FOR MEAT PRODUCTION

1 *Treat table birds as such from the start; do not name them.*

2 *Practise neck dislocation on a dead bird first.*

3 *Starve the designated birds for 12 hours (with access to water) to empty the gut.*

4 *Handle these live birds with care to avoid bruising.*

5 *Plucking is easier if the bird is still warm, then hang with the guts in.*

6 *Hang older birds in a fly-proof area for 2–3 days in warm weather, up to 7 days in winter, or to taste.*

7 *Maintain hygiene standards at all times.*

8 *Process just a few birds at a time to maintain enthusiasm and hygiene standards.*

9 *Skin casserole birds: this is quicker than plucking and the taste is not affected unduly.*

10 *Cook all poultry thoroughly.*

7

Diseases, problems and general troubleshooting

In this chapter you will learn:
- *how to cope with common diseases*
- *how to decide what is wrong with your bird*
- *how to treat your bird.*

Free-range poultry diseases

Symptoms	Disease	Cause	Treatment	Bird species
Listless, head sunk in neck, white diarrhoea, maybe blood in faeces	Coccidiosis	Coccidia parasite	Baycox, Coxi Plus or Coxoid* in water for five days. Keep litter dry	All birds from three weeks
Listless, head sunk in neck, yellow diarrhoea	Blackhead	Parasite carried by heterakis worm	Metronidazole in water for five days. Worm hens with Flubenvet to remove vector	Turkeys, pheasants, guinea fowl, uncommon in hens
White diarrhoea, thirst, sudden death	Bacilliary White Diarrhoea (BWD)	*Salmonella pullorum*	Blood test to find and cull carriers	Chicks zero to three weeks, adults as carriers
Listless, greenish diarrhoea, gaping, waterfowl lame	Parasitic worms	Up to six different types: round, tape, capillaria	Flubenvet mixed into feed for seven days, 10g to 8kg. Treat bi-annually	All poultry

Signs	Problem	Cause	Treatment	Birds affected
Visible parasites round vent, listless, blood spots on shell, dirty chest	Lice or mites	Four types louse, two types mite	Pyrethrum-based louse powder all over the bird, spray housing with Barricade or avermectin* drops on skin, 400 mcg/kg. Frontline*	All poultry. Care with avermectins in geese
Sneezing, nasal discharge, foam in eye, rattly breathing, swollen sinus	Mycoplasmosis	Mycoplasma	Tylan 200* injection, 0.5 ml per adult in breast muscle, 1 ml turkeys. Repeat 48 hrs. Tylan soluble for chicks	Hens, turkeys, peacocks, pheasants, ducks. Use Virkon to disinfect

* not licensed for poultry

(Contd)

Symptoms	Disease	Cause	Treatment	Bird species
Swollen sinus in waterfowl	Mycoplasmosis or pseudomonas	Mycoplasma or pseudomonas	Lance, flush with Baytril 2.5%, 5 ml teal, 30 ml swan, daily for five days. Spray fomites with Virkon	Waterfowl
Raised, encrusted scales on legs	Scaly leg	Mite	Dunk legs in surgical spirit weekly for three weeks or avermectin* drops on skin. Scales take one year to grow back properly	Any bird

Blood	Wounds	Feather pecking: heat stress, overcrowding	Remove red colour, spray coloured antiseptic, put Stockholm tar on area. Isolate till healed	Any bird
Brown diarrhoea, slow growth, slow feathering	Enteritis	E. coli, stress, dirty litter, chilling	Terramycin or Apralan or Baytril in water	Young stock from five days
Noxious smell, scabby vent	Vent gleet	Herpes virus	Acyclovir* may help but best to cull	Birds over one year
Purple comb when normally bright red	Heart failure, nitrate poisoning	Age, disease, deformity	No treatment if heart, methylene blue if nitrate poisoning	Hens
Swelling on underside of foot	Bumblefoot	Staph bacteria, wound	Surgical intervention, sulphur/silicea tablets	Old, heavy birds, perches too high

* not licensed for poultry

(Contd)

Symptoms	Disease	Cause	Treatment	Bird species
Top beak overgrown, long claws	Overgrowth	Deformity, soft ground	Trim to shape	All birds
Female flank no feathers or bleeding	Bareback	Sharp claws or spurs	Trim to shape, use saddle for breeding	Soft feathered hens, turkeys
Unusual behaviour	Stress	Disturbance or changes	Vitamin powder, probiotics	Any bird
Lameness (if waterfowl, also see 'worms')	Injury, tumour	Kidney disease, tumour, arthritis, perosis	Isolate. Symptomatic treatment, NSAIDs*, for perosis, cull and check Manganese in feed	Any bird
Paralysis, same side leg and wing	Marek's disease	Herpes virus	Vaccinate or cull affected birds. Keep chicks and adults separate	Hens up to point-of-lay, cockerels

Symptoms	Name	Cause	Treatment	Birds affected
Weight loss but still feeding and alert	Avian TB	*Mycobacterium avium*	No treatment. Cull. Wild birds are carriers	Birds over one year
Respiratory distress, loss of egg quality	Infectious Bronchitis (IB)	Coronavirus	Vaccinate. Carrier adults, 40 per cent chick mortality	Hens, pheasants, guinea fowl
Respiratory distress, gasping, death	Aspergillosis	Fungus	Cull, remove damp litter, fog with F10	Zero to four weeks chicks, turkeys, water fowl pheasants plus any adults
Pendulous crop	Cropbound	Fibrous grass, poor muscle tone	Isolate with only water 48 hrs. Surgery if impacted	Old hens or birds on long grass/hay
Sudden chick deaths, rest recover	Gumboro (IBD)	Birnavirus	Vaccinate, antibiotic cover useful	Hens one to 16 weeks

* not licensed for poultry

(Contd)

Symptoms	Disease	Cause	Treatment	Bird species
Cheesy substance in ear canal	Ear infection	Bacteria or mites	Surolan*, Leo Yellow tube*, Aurizon*	Any bird
Cheesy substance in mouth and throat	Canker	Trichomonas	Metronidazole in water. Vitamin A	Any bird
Listless, straining	Egg bound	Low available calcium or stress	Cull if egg peritonitis. Warmth, remove egg, give Calcivet or Nutrobal	Any female bird
Waterfowl wing droops then stricks out	Angel wing	High protein diet	Drop protein, tape wing in natural position for three days, do again if necessary	Growing waterfowl

Waterfowl no longer waterproof	Wet feather	Shaft lice, excess preening, tree mould	Keep off water until next moult, de-louse, if mould, wash in detergent	Waterfowl
Sudden death breeding waterfowl	Duck viral enteritis	Herpes virus	Vaccinate. Remove wild mallard	Waterfowl

* not licensed for poultry

Common problems and some causes

WEIGHT LOSS

Liver disease; starvation; bullying; avian TB; Northern fowl mite or red mite; poisoning, coccidiosis; kidney disease; lack of water; high levels of ammonia.

DIARRHOEA

E. coli; Bacilliary White Diarrhoea (BWD); coccidia; too much cabbage; hexamita; *Salmonella typhimurium* or *enteritidis*; sudden change of diet. Do not confuse caecal contents (1 in 10 voiding) with diarrhoea.

MISSING FEATHERS

Moulting; pushing head through fence; hens plucking each other or cockerel (culprit usually has all feathers); claws of male; poor nutrition.

NO EGGS

Birds too young; days too short; Infectious Bronchitis (IB); fright/stress; new home; rats or magpies stealing eggs; birds laying away; eggs laid on floor and buried; eaten by hens; not enough food; too much food; Northern fowl mite or red mite.

CHICK DISCOMFORT

Noisy when defecating and pasted-up vent: E. coli or chilling. Continual cheeping: too cold, hungry, thirsty; one escaped from brooder. Panting: too hot. Huddled together: too cold.

RESPIRATORY NOISES

Mycoplasma; IB; aspergillosis; high levels of ammonia; gapeworm.

EGGS

Infertile: excess males; flea eggs on vent; too many feathers on vent; cock too old. Not hatching: not knowing correct incubation time; not fertile; infected by hen. Dead-in-shell: dirty shell/nest; 'banger' (egg explodes in incubator as rotten), too much humidity in setter; not enough humidity in hatcher; sat on by broody before collection; poor nutrition of breeders; poor egg storage; eggs too old (> 14 days); old age of hen; drugs at incorrect levels; *Salmonella pullorum*; infection, damaged shell, malposition of embryo.

CHICKS

Early hatch: very small bantams, eggs too fresh (< 24 hours); sat on by broody before collection. Unhealed navel: eggs too old; temperature of incubator too low; too much humidity in setter. Deformities: genetic or nutritional; slippery surface for first few days. Deaths: salmonellosis; coccidiosis; chilling; smothering.

SUDDEN DEATH IN ADULTS

Egg peritonitis; heart failure; Gumboro; salmonellosis; stoat/mink/ferret/fox; choked; kidney failure; aspergillosis; botulism; duck viral enteritisi Newcastle Disease.

Common diseases by age

CHICKS

Deformities (bent toes, crossed beak, splay leg); *E. coli*; BWD (*Salmonella pullorum*); Gumboro; IB; coccidiosis, starve-outs; feather pecking. Natural rearing includes the previous conditions plus chilling, squashing, vermin.

GROWERS (8–26 WEEKS)

Northern fowl mite/red mite; scaly leg; coccidiosis; mycoplasmosis; swollen sinus; feather pecking; angel wing; perosis; smothering;

Marek's; lameness; roach back; wry tail; cow hocks; impacted gizzard; roundworms; breathing difficulty (IB, gapeworm); poisoning; blackhead; aspergillosis; lice.

ADULTS AND AGED

Egg peritonitis; avian TB; heart failure; scaly leg; Northern fowl mite/red mite; lice; mycoplasmosis; swollen sinus; ear canal infection; bumblefoot; vent gleet; wet feather; arthritis; choking; sour crop; impacted crop; impacted gizzard; tumours; poisoning; aspergillosis; kidney failure; liver failure.

Life expectancy

Large fowl, turkeys and guinea fowl: six to ten years. Bantams, peacocks and pheasants: eight to twelve years. Ducks: six to eight years. Geese: ten to twenty years. Quail: one to two years.

Pages 202–209 comprise a poster sponsored by the British Veterinary Association (www.bva-awf.org.uk) and the Poultry Club (www.poultryclub.org). The information is taken from *Diseases of Free-Range Poultry* by Victoria Roberts BVSc MRCVS, published by Whittet Books (2009), by kind permission of the publisher.

Description of major diseases

CONTROL OF COCCIDIOSIS IN CHICKENS

Cause
This is a protozoal parasite that multiplies in the gut and is specific to different hosts. Not all species of coccidia are harmful but there are seven of the *Eimeria* species pathogenic to chickens.

Life cycle
▶ *Parasitic phase: the infective oocyst (coccidia egg) is eaten by the chicken and then multiplies over about seven days within the gut of the bird, millions of new oocysts resulting from just one ingested oocyst.*
▶ *Non-parasitic phase: excreted in the droppings, the oocysts then take two days to mature before being ready for the next host to eat.*

Progression
Dayold chicks do not get immunity from their mother. Birds of any age are susceptible, but most acquire infection early in life, which gives them some immunity.

Immunity is best kept strong by a low level of infection, which is what happens on free-range.

Birds kept or reared on litter are more at risk as the coccidia has conditions that suit it, such as wet litter. If the birds are also stressed by environmental factors (cold, overcrowding, poor ventilation) then disease results. The oocysts are very resistant to destruction, either by disinfectants or by drying out.

Symptoms
The seven species of coccidia prefer different areas of the gut, some producing the expected bloody diarrhoea, some producing high levels of mucus, sometimes white diarrhoea, and others stunting growth. Infection can show from three to six weeks of age and infective oocysts can be transported by people looking after the birds. Older birds can become infected if either their immunity has been reduced due to being kept on a wire floor (no access to droppings and therefore no trickle infection) and then put onto litter, or if environmental stressors reduce their immunity. The birds generally look hunched and depressed with or without blood in the droppings.

Prevention
Good biosecurity is important. Anticoccidial drugs in the feed for the first six weeks of life has been the norm in order to let

the chicks have a low level of infection and therefore acquire immunity. With the reduction of permitted drugs in feed, the vaccine using weakened coccidia, which has been used by commercial units for over 14 years, is the way forward. Free-range reduces the incidence of disease whilst still providing trickle infection to boost immunity. Oocysts can stay in sheds despite disinfection unless a specific oocidal ('egg-killing') disinfectant is used.

Treatment

The only product currently licensed for treatment in chickens is Baycox and use is restricted to broiler breeders on the data sheet (licence details). Sulphonamides can be ordered from specialist poultry veterinary practices. On a small scale and where the chickens do not enter the food chain, the pigeon product, Coxoid, is used by fanciers. This contains amprol, which used to be licensed for chickens and has proved to be safe.

By far the better treatment and prevention is the vaccine, Paracox. This contains all seven species of coccidia but these are weakened so that they cause the chicken to mount an immune response but not to become infected. As this is an industrial product it normally comes in quantities to treat thousands of birds. It is now available in small quantities to *bona fide* Poultry Club members via the website (www.poultryclub.org) and through order forms. Paracox is administered once as a solution from a dropper bottle to a healthy dayold chick via its mouth. The shelf-life of the product is four weeks, so orders need to be made with the monthly expected hatch in mind. Any feed used for vaccinated birds should not contain anticoccidial drugs as this will counteract the vaccine. The vaccine can be used on unvaccinated chicks up to nine days old but is most effective at dayold.

Other species of poultry

Young turkeys and young waterfowl are susceptible to their own species of coccidia. There are no products licensed for treatment in turkeys and waterfowl but Coxoid has been used with success. The coccidia life cycle is similar to that of chickens and the oocysts persist in the ground.

CONTROL OF MAREK'S DISEASE IN CHICKENS

Cause
A herpes virus that both gravitates towards lymphoid tissue (lymphoid tissue is spread throughout a chicken, unlike a mammal where it is confined mostly to lymph nodes) and also causes demyelination of peripheral nerves (leg and wing). Various strains of the virus have been isolated and there are two types, one that lives in the cell and one that is cell-free and lives in the feather follicle. The virus was discovered in 1907 by Jozsef Marek, a Hungarian vet, and there are non-pathogenic, mildly virulent, virulent and very virulent strains.

Life cycle
The virus replicates in the host chicken's lymphoid tissue and is shed in feather dander. The virus can remain viable for at least one year in feather dander and henhouse dust.

Progression
Chickens from six weeks of age are affected and symptoms are most frequently seen at 12–24 weeks of age with the hormonal stress of point-of-lay (POL) being a classic time for the signs to appear. Older birds can sometimes be affected if stressors (changes in weather, food, handling, environment) are not minimized. Females are more susceptible than males.

Symptoms
Mortality is variable and depends on which of the peripheral nerves is affected, but leads to progressive spastic paralysis of the legs and wings. Sometimes, if the neck nerves are affected, the neck can twist around. There is an acute form where birds may die suddenly with no symptoms and tumours may be found in the liver, gonads, spleen, kidneys, lungs, proventriculus, heart, muscle and skin.

Infection

The virus is ubiquitous in poultry worldwide. Infection in chicks occurs by inhalation and two weeks later, the virus is shed by the infected chick in feather dander and by oral and nasal secretions. The virus is *not* passed on through the egg. Infected and recovered birds continue to shed the virus for life.

Prevention

Good biosecurity is important, so quarantine any new stock for four weeks. Rear chicks for two to three months away from adult feather dander if adult birds have shown symptoms. Ask vendors if stock has been vaccinated. Genetically resistant breeds include the Fayomi.

Treatment (control)

Cull any affected birds. This increases the resistance to the disease in the surviving birds. Vaccination is feasible, especially if Silkies or Sebrights are kept. These are very susceptible to clinical signs of Marek's and there would be few of these breeds seen at exhibitions if vaccination was not used. The vaccine is administered once by injection, ideally when chicks are dayold or before three weeks old.

In other breeds, using vaccine can hide the virus and so the whole stock gets progressively more susceptible (weaker) without any symptoms and if birds are sold without the recipient being told of the vaccination, the birds can pass on the virus to unvaccinated chicks, thereby bringing the disease to a flock that may have been free of it before.

Diagnosis

Clinical signs of paralysis and postmortem lesions.

MYCOPLASMA IN POULTRY: WHAT IT IS AND WHAT TO DO ABOUT IT

Introduction

Mycoplasma in poultry is not a new disease. There is mention in the old books of similar symptoms from about 100 years ago but it has generally been called roup or a common cold. Treatment tended to be by culling only.

The disease acquired the name mycoplasma once the causative organism had been discovered. Mainly the respiratory system in poultry is affected and the disease may be becoming more common, spreading with increased travelling of stock, or it may be that we are hearing about it more with improved communications. The incubation period before clinical signs appear can be as little as a few days – it is very infectious. It appears to thrive in the bird when other pathogens are present, such as *E. coli* or IB (IB is certainly now more common in free-range flocks) or if the birds are stressed or debilitated. Debilitating factors include nutritional deficiency, excessive environmental ammonia and dust, and stressors such as changes in the pecking order or exhibitions.

Cause and clinical signs
The organism is neither a bacterium nor a virus in size, but partway between, having no cell wall but with a plasma membrane. Four out of the known 17 species of mycoplasma are pathogenic in poultry:

▶ *Mycoplasma gallisepticum: signs can include foamy eyes, sneezing, nasal discharge, swollen eyelids and sinuses, reduced egg production and gasping in chickens, turkeys and pheasants, swollen sinuses in waterfowl. This one is the main culprit in backyard flocks.*
▶ *Mycoplasma synoviae: signs include swollen and hot joints in chickens and turkeys and/or respiratory signs as above.*
▶ *Mycoplasma meleagridis: signs include poor growth in turkey poults and lowered hatchability in turkey breeders.*
▶ *Mycoplasma iowae: signs include reduced hatchability in turkey breeders and twisted legs in turkey poults.*

When nasal discharge is evident, feathers become stained with this as the bird tries to clean its eyes and nostrils. There is a particular sweet smell associated with this discharge which is immediately apparent to the sensitive nose when entering a henhouse.

Transmission

Nasal discharge and cool temperatures are protective of
the organism so any sneezing will deposit droplets that will
remain infective for several days. Transmission is also through
the egg, plus carried on the clothes and hands of people tending
the birds.

Treatment

Antibiotic treatment will not completely cure the disease but
will reduce the incidence to a tolerably low level. Tylan Soluble
is licensed for the treatment of mycoplasma, as is Baytril. These
oral preparations are effective in young stock but seem to be
less effective in older stock. Tylan 200 injection (not licensed for
poultry) is effective with 0.5 ml in the breast muscle of an adult
large fowl, repeated 48 hours later if still sneezing. If still noisy
after that the bird must be culled as the organism will be too deeply
entrenched within the airsacs and hollow bones to be removed, the
bird remaining a carrier that will infect others. The reason Tylan
200 is not licensed for poultry is because it harms muscle, which in
a meat bird is disastrous but in backyard or fancy poultry that do
not enter the food chain, it is not really an issue.

Prevention

- *Keep stressors to a minimum or if a known stressor such as
  a show is imminent, give vitamin supplementation. There are
  several useful products on the market that contain probiotics
  and vitamins, administered in the water.*
- *Use a suitable disinfectant for both huts and equipment, such
  as Virkon.*
- *Keep dust and ammonia levels low. Ammonia paralyzes the
  small hairs that act like an escalator to move normal mucus up
  the trachea before being swallowed.*
- *Feed high quality commercial food for the stage of growth and
  the species of bird.*
- *Monitor weather changes and take steps to minimize
  any effects.*
- *When attending to the stock, begin with the youngest at the
  start of the day (i.e. with clean clothes).*

- *Either quarantine new stock for two to three weeks or inject once with Tylan 200 as soon as the birds are obtained if there has been mycoplasma in your flock.*
- *Some very conscientious breeders inject stock they sell and warn buyers of the disease risk.*
- *Do not buy from auctions.*
- *If adult stock are kept symptom-free the risk of passing mycoplasma on through the egg is reduced.*
- *If young stock happen to be exposed to a mild bout of mycoplasma they will acquire a certain amount of immunity as long as there are no other pathogens present.*
- *Biosecurity.*

Vaccination

There is a mycoplasma vaccine marketed by Intervet but it is not recommended for use in breeding birds. This appears to be because the manufacturers do not know how long the vaccine is effective.

With vigilance, mycoplasma can be kept at a low level in backyard flocks thus increasing the welfare of the birds.

INFECTIOUS BRONCHITIS

IB is a viral disease initially affecting the respiratory system in a young bird but can localize in the reproductive tract and cause misshapen or soft–shelled eggs with very watery whites when the bird begins to lay. There is a vaccine for this, but no treatment if eggs are affected.

10 THINGS TO REMEMBER FOR DISEASES

1 *Check if birds you are buying have been vaccinated.*

2 *Do not buy from auctions unless you know the seller.*

3 *Quarantine new birds for three weeks.*

4 *Check birds visually every day for all the normal signs (see relevant chapter).*

5 *Check birds by handling once weekly.*

6 *Check colour and consistency of droppings daily.*

7 *Contact a vet when you first get the birds so that when you need them, they know who you are and what you have got.*

8 *At the first sign of snivelling, get the birds treated.*

9 *Treat against worms at least twice yearly, more often if on small area.*

10 *If breeding chickens, vaccinate the dayolds with Paracox.*

8

Cooking with eggs

In this chapter you will learn:
- *some good ways of cooking with your own eggs*
- *delicious new recipes.*

Favourite recipes

Mrs Beeton in 1861 extolled the virtues of eggs from the common hen as being most esteemed as delicate food, particularly when 'newlaid'. William Cobbett said stale eggs are things to be run from, not after. If you have a glut of eggs, they can be successfully frozen if first separated and then subsequently used in cooking. Add one duck egg to cake recipes for the most moist and delectable cakes – prize-winning!

Fresh eggs do need to be treated with respect and delicacy. Some people keep eggs in the fridge, influenced probably by the egg-shaped storage areas seemingly in every one. Duck eggs need to be kept at around 4° C (39° F), but it is sufficient to keep other eggs below 10° C (50° F), which usually means a larder in an older house.

Insight

Eggs that have been kept in the fridge need to be brought to room temperature before using, so take them out a couple of hours before you need them. You will get better results.

Meringues are a disaster using fresh eggs, the white is just too moist and will not peak, so you need to plan for meringues by keeping enough whole eggs back for about ten days and then use them.

When you have had your own eggs for a while, the time will come when an egg appears that may have been hiding in the litter and even sat on by an undetecting broody (at night). After ruining a complete recipe with one of these, it will become a habit to shake each egg beside your ear. If it rattles, bin it.

Commercially produced eggs are machine candled to weed out those with defects. Only feed cooked eggs to dogs as raw ones bind biotin in the canine gut, making their claws brittle.

FINEST BOILED EGGS

▶ Boil enough water to cover the eggs in a pan. When the water is boiling, remove the pan from the heat and very carefully lower the required number of chicken or turkey eggs into the hot water. Wait a moment for air bubbles to rise then return the pan to the heat and put on simmer for five to six minutes or to taste. This prevents fresh eggs from cracking.
▶ If boiled eggs are needed for a picnic or lunchbox, immerse them in cold water once cooked, for five minutes.
▶ Peel (with difficulty if fresh) and dip into celery salt.

FINEST POACHED EGGS

▶ Heat some water in a shallow pan until boiling.
▶ Remove the pan from heat and crack the required number of chicken or turkey eggs carefully into the hot water.

> ▶ *Return the pan to a simmer heat and flick water over eggs with a spatula to cook the surface. This makes fresh eggs keep their shape.*

FINEST BREAKFAST OR SUPPER SCRAMBLED EGGS

butter and olive oil
3 chicken eggs or 2 duck eggs or 1 goose egg per person, beaten
freshly ground black pepper to taste
1 tbsp hot water added to beaten egg immediately before putting
in pan (never milk as it makes the eggs hard and dry)
oak-smoked salmon slices

> ▶ *Heat a little butter and olive oil in a heavy or non-stick pan until hot, then pour in the beaten eggs.*
> ▶ *Stir continuously, reducing the heat, until cooked to taste. A teaspoon of cream per person for special occasions can be added here.*
> ▶ *Season to taste and serve immediately with oak-smoked salmon slices (these should provide enough salt).*

INDIAN BREAKFAST OMELETTE

butter or olive oil
2 eggs, beaten
3 cherry tomatoes, quartered
1 spring onion, chopped
2 small green chillies, chopped
2 medium-sized mushrooms, chopped
1 tbsp grated or crumbled Cheddar or Lancashire cheese (optional)

> ▶ *Heat a 25 cm (10") omelette pan and coat with butter or olive oil.*
> ▶ *When the pan is hot add all the quartered and chopped ingredients (and cheese if using) to the beaten eggs so they are coated.*
> ▶ *Pour the mixture into the saucepan, gently making breaks in the omelette with a spatula so that the raw egg runs through*

and cooks; keep going round the edge with a spatula to keep the omelette a good shape and prevent sticking. Cook to taste and fold onto a plate.

CRUMPET BREAKFAST

This makes a jolly good supper dish after a hard day tending hens.

- ▶ Make a cheese sauce (the simple warmed double cream with melted cheese version is far superior to the white sauce with cheese added version).
- ▶ Take an equal number of crumpets and eggs (one or two per person).
- ▶ Take oodles of spinach (or spinach beet, which has less oxalic acid and is really easy to grow).
- ▶ Toast the crumpets.
- ▶ Poach the eggs.
- ▶ Cook and chop the spinach.
- ▶ Stack: crumpet topped with chopped spinach topped with poached egg topped with cheese sauce.

TARRAGON BAKED EGGS

butter for greasing
2 tbsp double cream
1 tsp fresh tarragon, chopped
1 fresh egg
sea salt and freshly ground black pepper

- ▶ Preheat the oven to 190° C, Gas mark 5.
- ▶ Grease a ramekin and add 1 tbsp cream and half the tarragon.
- ▶ Put the ramekin in a roasting tin and pour boiling water into the tin to half way up the ramekin.
- ▶ Set over moderate heat until the cream is hot.
- ▶ Break the egg into the ramekin (carefully), season then cover with the remaining cream and herbs.
- ▶ Put the tin in the oven and bake for seven minutes until the yolk is just set.

KEDGEREE

Traditionally this is a breakfast dish but it is also a good supper dish with the addition of a salad.

85 g (3 oz) cooked brown rice per person
1 hardboiled egg, coarsely chopped, per person
60 g (2 oz) flaked oak-smoked haddock per person
30 g (1 oz) butter per person
sea salt and fresh ground black pepper to taste

▶ *Preheat the oven to 190° C, Gas mark 5.*
▶ *Mix all the ingredients gently.*
▶ *Place in an ovenproof dish covered in foil and heat in the oven for 15 minutes.*
▶ *Serve with fresh brown toast.*

BEBINCA (LAYERED DESSERT)

1 cup coconut milk
400 g (14 oz) granulated sugar
10 eggs, beaten lightly
1 cup refined plain flour
½ tsp salt
¼ tsp nutmeg
125 g (4½ oz) ghee, melted

▶ *Preheat the oven to 180° C, Gas mark 4.*
▶ *Dissolve the sugar in the coconut milk in a pan over a low heat. Keep aside to cool.*
▶ *Mix in the beaten egg a little at a time.*
▶ *Gradually add the remaining ingredients except the ghee. Mix the batter until smooth, then strain.*
▶ *Pour 3 tbsp of warmed ghee into an 18-cm round, deep baking tin. Pour 1 cup of batter over the ghee and bake for ten minutes until set and the top is golden.*
▶ *Pour 1 tbsp of warmed ghee over the baked layer and then ¾ cup of batter.*

- ▶ *Bake till golden.*
- ▶ *Continue until all the batter is used up. There should be seven layers.*
- ▶ *When cool, turn out the bebinca onto a plate. Leave to set for at least 12 hours before slicing.*
- ▶ *Serve with ice cream or cream.*

CHOCOLATE RASPBERRY ROULADE

8 eggs (separated)
1½ cups sugar
¼ cup bittersweet chocolate
2 tsp butter
vanilla extract
butter, melted
whipping cream
sugar
red raspberries

- ▶ *Gently crack one egg on a flat surface and break the shell into two pieces.*
- ▶ *Pull the pieces apart, keeping the yolk in one half and allowing the whites to fall into a bowl.*
- ▶ *Transfer the yolk to the other egg shell-half to allow the rest of the whites to fall into the bowl.*
- ▶ *Place the yolk in a separate bowl.*
- ▶ *Follow this procedure until all the eggs have been separated.*
- ▶ *Add 1 cup of sugar to the bowl with the egg yolks. Whisk until mixture is light and fluffy and has a ribbony consistency. This will take about three to four minutes.*
- ▶ *Chop the chocolate into smaller pieces. Place the butter and chopped chocolate into a saucepan and melt over a low heat. Stir frequently with a wooden spoon.*
- ▶ *While the chocolate is melting beat the egg whites until they form soft peaks.*
- ▶ *Stop beating and add half a cup of sugar. Continue beating until they form stiff peaks.*
- ▶ *Combine the melted chocolate with the egg yolk mixture.*

- ▶ Add a few drops of vanilla extract. Mix until well combined.
- ▶ Stir a quarter of the beaten egg whites into the chocolate mixture.
- ▶ Gently fold in the remaining egg whites until completely mixed.
- ▶ Carefully pour the cake mixture into a 35 × 25 cm (14" × 10") rimmed, non-stick baking sheet. Make sure the mix is evenly distributed.
- ▶ Bake for 15 minutes in a oven set at 180°C, Gas mark 4.
- ▶ While the cake is baking prepare the whipped cream. Add very cold cream to a chilled mixing bowl. Whip the cream to soft peaks.
- ▶ Stop whipping and add a small amount of sugar. Continue to whip until stiff peaks are formed.
- ▶ Remove the cake from the oven. Let it cool for ten minutes in the baking sheet on a rack.
- ▶ Turn out of the baking sheet onto a waxed paper lined rack.
- ▶ Let it cool completely.

For individual roulades
- ▶ Cut the cake with a serrated knife lengthwise into three equal strips of just over 8 cm (3") width.
- ▶ Cut each strip widthwise into three pieces of about 11 cm (4") length.
- ▶ Spread each piece with whipped cream, leaving a 1.5 cm (½") border all the way around.
- ▶ Sprinkle with raspberries and roll up gently.
- ▶ Clean edges if any cream comes out.
- ▶ Refrigerate until ready to serve.

For one large roulade
- ▶ Spread the cake with the cream, leaving a 1.5 cm (½") border all the way around.
- ▶ Sprinkle with raspberries.
- ▶ Starting at the short end, roll up gently.
- ▶ Refrigerate until ready to serve.
- ▶ To serve, cut into 2.5 cm (1") widths.

MUSTARD CHILLI EGGS

This can be scrambled and served on a bagel for breakfast as well.

2 extra-large eggs
dribble of water
¼ tsp minced chilli
¼ tsp wholegrain mustard
1 tbsp butter for cooking

- ▶ *Break the eggs into a bowl and beat thoroughly.*
- ▶ *Add a little water and continue beating until the mixture becomes quite frothy.*
- ▶ *Add the chilli and mustard and continue beating the mixture until thoroughly mixed.*
- ▶ *Melt the butter in a 25 cm (10") frying pan over a medium heat.*
- ▶ *While the butter is melting and heating up to a sizzling stage, beat the egg mixture constantly.*
- ▶ *Add the mixture to the pan and hear it sizzle.*
- ▶ *Shake the pan occasionally and leak the uncooked egg mixture around the edges till the omelette is almost thoroughly cooked.*
- ▶ *Fold the omelette in the pan using a spatula.*
- ▶ *Serve immediately.*

Appendix 1: Vermin control

Where there is livestock and their food there will always be vermin, so prevention is the best form of protection. Protect yourself from disease by wearing gloves when handling dead mice and rats.

Types of vermin

- ▶ *Mice are attracted by food and can get in via very small holes.*
- ▶ *Rats are attracted by food and eggs. Poison or trap them.*
- ▶ *Grey squirrels are attracted by food and eggs and are worse in urban areas. Poison or trap them.*
- ▶ *Weasels or stoats are attracted by the birds. Trap or prevent access to poultry houses.*
- ▶ *Mink are attracted by the birds. Trap or prevent access to poultry houses.*
- ▶ *Foxes are attracted by the birds. Prevent access to poultry houses.*
- ▶ *Feral cats are attracted by the birds. Prevent access to poultry houses.*
- ▶ *Magpies, crows, jackdaws and rooks are attracted by food and eggs. Prevent access to poultry houses.*

Animals protected by law are owls, herons and hedgehogs.

Live catch traps are useful in case the wrong species is trapped but then there is the problem of despatching the beast. This must be done humanely and an air rifle at close range is probably the most accessible means. Rat traps can also be the breakback type and must be set in a tunnel so that other species are excluded. Always wear gloves both to disguise human scent and to avoid Weil's disease (leptospirosis) that rats commonly carry. Multicatch mouse traps can be placed inside the henhouse, but remember to check and empty them daily as the mice tend to die overnight.

Feed and water should be inside the henhouse or a covered run. Feed should be stored in metal or plastic bins. The only open access to a henhouse during the day should be a pophole and if magpies get through this, pin vertical strips of black binbag over it – the hens will push their way in but the magpies will not like the movement of the strips.

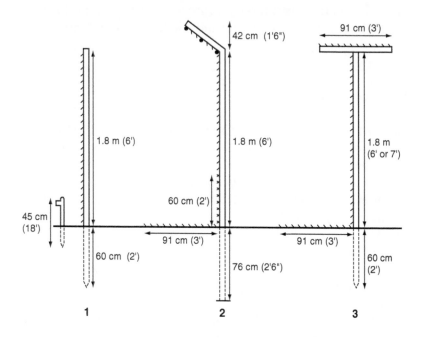

Figure Ap1.1 Three styles of fox-proof fencing.

Plastic electrified netting 1.2 m (4″) high such as Flexinet is quite useful for poultry, keeping the birds to a designated area, as long as the pulse unit is powerful. It is not 100% fox-proof and can be dangerous if waterfowl get stuck in it, which they have a habit of doing.

Poison

Keep permanent poison bait areas organized; protect them from pets and children and preferably use the newer poisons that act

230

only on rats' digestion (e.g. Eradirat®), not the anticoagulants, which can affect non-target species. Despite their cute antics in gardens, grey squirrels are a huge nuisance by taking food and eggs. They will eat (warfarin) poisoned maize but this must be placed in a special labelled device, again, to avoid targeting the wrong species. They have a vicious bite, so wear gloves.

Access can be prevented for foxes, cats, mink etc. by the use of fencing or covered runs.

Appendix 2: Transport recommendations

▶ *Cardboard boxes can be used for transporting poultry. Use stout ones and create ventilation holes by making two parallel cuts about 2.5 cm (1") apart across at least two corners and push the centre section inwards. 'Weave' the top so that it is secure and then tie with string like a parcel. Discard after single use.*

▶ *Alternatively, make ventilated wooden boxes to suit the size of bird but varnish them so they can be disinfected.*

▶ *Boxes should be placed on the back seat of a saloon car and not in the boot. Estate cars, saloons and vans should have sufficient ventilation by opening windows or the use of air conditioning.*

▶ *A plastic poultry crate is ideal for transporting birds in numbers as it is easy to clean and disinfect. It is also airy and food and water containers can be attached easily. If a trailer is used for transport, make sure there is adequate ventilation for the birds both when travelling and when static.*

▶ *Food and water must be provided for journeys over 12 hours.*

▶ *Fill in and carry with you a transport declaration certificate form (www.poultryclub.org) for journeys outside your local authority area.*

Appendix 3: Exhibiting and show preparation

This does not just mean having clean birds. Preparation starts months before a show because fitness (correct feeding for good bone and muscle) is the framework upon which all the superficial items such as feathers are built. With some early maturing breeds it is only possible to show them in their first year, which usually means just coming into lay, so that hatching has to be timed to match point-of-lay (POL) (around 18 weeks) and the chosen show(s). This effectively means that if these breeds do not have all the breed points when they are young, they are unlikely ever to have them. Other breeds do not mature until they are at least two years old, so will gain in breadth of body if the frame is there as youngsters. Ducks with mallard–type colouring can be a problem as they go into eclipse during the summer months and do not regain their breeding plumage until early October.

Even dark-coloured birds need to be washed for a show. Either washing-up liquid or baby shampoo is normally used. The birds are dunked in warm water, lathered, rinsed and dried either in front of a fire or with a hair drier. They rather like this. Don't forget the legs, which may need to be scrubbed gently with an old toothbrush; also ingrained dirt will need to be gently removed from under the scales with a wooden toothpick. Waterfowl legs may need to be scrubbed gently, taking care not to rub the sensitive bill too hard as the top cuticle containing some colour may be removed.

Put waterfowl and turkeys in a stable with clean shavings on the floor if the outside temperature is not too low. It is best to wash birds at least a week before a show to allow the natural body oils to return to the feathers. Put the birds in a clean show pen in an area with lots of human activity to get them used to the bustle of a show. If the birds are tame as well, then so much the better.

Poultry with feather colours that are liable to fade or change in strong sunshine tend to be kept in outdoor runs that are covered over, that means they are still fit and still the correct colour. If you must wash a bird the day before a show, make sure it is dry before you box it, otherwise the feathers will stick out at all angles. Always try to use boxes that are too big so that the birds have enough room both to keep cool and to turn around, which protects tail and wing feathers.

Use a proprietary louse powder to make sure that no unwelcome parasites accompany your birds to the show. You will have taken precautions against scaly leg mite, for instance, by dunking the legs in surgical spirit about once a month throughout the year, so that should not be a problem. Make sure that claws and beak are trimmed to shape. Dog toenail clippers are useful for this.

Check on the correct leg colour for your breed, because if it should be yellow, and the birds have been laying well, the yellow colour will go out of the legs into the yolks. Feeding some maize will help to counteract this, as will running the birds on grass all the time. By the same token, if your breed should have white legs, do not feed maize in order to avoid a yellow tinge to the skin. Grass does not turn white skin yellow.

All shows have an entry date, which varies between several weeks to one week before the show. Make sure you enter before this date as late entries are not accepted and check that your entries are correct for the various classes. Show Secretaries will give entry information if asked. Their addresses are in the Poultry Club *Yearbook* under Affiliated Societies, and lists of shows are usually published in the various poultry magazines. If you have shown the previous year you will normally be sent a schedule for that show. Make sure that your birds are penned in time for judging, and a little oil or vaseline rubbed on the comb, wattles and legs will spruce them up. A silk handerchief is good for imparting a shine to the feathers, but it is more enduring to have the shine there through good feeding and management in previous months.

Birds are not normally fed or watered in show pens before judging as this can change the correct outline or create dirt and droppings, but take food in the form of grain (firmer droppings) to a show plus water in a container suitable to pour through the bars of a showpen after judging, as not all shows are of sufficient duration to afford stewards time to feed and water birds. Water containers are usually provided, but if in doubt, take a container that can be wired, pegged or fixed to the pen so that it does not tip over. It is your responsibility to check that your birds have been fed and watered.

Bear in mind when returning from a show that dusting all birds with louse powder is a sensible precaution, and ideally, all show birds should be kept separate from your other stock for a few days just to make sure that they have not brought something contagious home from the show, or that the stress of showing has not depressed their immune system, allowing the entry of disease. A bit of cosseting after a show may well mean that a particular bird can be shown again soon, or return to the breeding pen in a fit condition.

Ensure you comply with all biosecurity regulations.

Appendix 4: Update on avian influenza and registering of poultry flocks

Key points

- *Avian influenza (AI) is a disease of birds, not humans. However, humans can, rarely, be affected.*
- *There are both high pathogenic (HPAI) and low pathogenic (LPAI) forms and many strains.*
- *LPAI does not always show up as disease in birds. However, it is present in some areas of the global wildfowl population.*
- *There is a constant but low risk of migrating birds bringing LPAI to the UK.*
- *LPAI can mutate into HPAI, especially when introduced into poultry populations.*
- *Some strains of HPAI spread easily between birds and cause illness – with a high death rate – very quickly in poultry populations.*
- *In rare cases, some HPAI strains can lead to severe illness and deaths in humans where there has been close contact with infected birds.*
- *There are a limited number of reported cases of human–to–human spread of AI. There is no such thing as a human pandemic of bird 'flu.*
- *Migratory waterfowl – particularly wild ducks – are the main known natural reservoir of AI viruses. These birds are the most resistant to clinical disease and often show no clinical signs or mortality when infected.*
- *It is possible for human and bird 'flu viruses to combine to produce a new human 'flu virus if a person is infected by both at the same time. This could produce a virus to which people have no current immunity and that could spread between humans.*

- It is therefore very important to ensure that any outbreak of AI is controlled quickly and that workers and veterinarians in close contact with infected birds are well protected. The UK government has contingency plans in place to ensure that this is so.
- The UK government and key stakeholders are working closely together to ensure that the UK response to current circumstances is appropriate and comprehensive.
- Advice on worker protection is a matter for the Health Protection Agency.
- Keepers of poultry will wish to be vigilant, to take care if handling birds that appear to be unwell and to observe high levels of biosecurity. Guidance is on the DEFRA (UK Department for Environment, Food and Rural Affairs) website.
- In the event of an outbreak, DEFRA will be able to protect its workers with the necessary drugs and equipment. Detailed instructions to staff are in place.
- Surveillance is important in order that arrival of AI in the UK is rapidly detected.
- The public is encouraged to report (not touch) 'dramatic local incidents' – large numbers of sick, dying or dead birds to the DEFRA hotline 08459 335577.

Following a ban on poultry shows, sales and gatherings at the end of October 2005 as instigated by the European Union (EU) as a precaution against the risk of AI, DEFRA announced just before Christmas 2005 that poultry shows and sales would be allowed in the UK under a general licence and if they conformed to certain biosecurity conditions. Flocks of over 50 must have been registered with DEFRA by 28 February 2006, while voluntary registration of smaller flocks continues after this date. Details are below, but visit the DEFRA website (www.defra.gov.uk) for more information.

Avian influenza (bird flu) – schedule of general licence conditions

The following licence conditions will apply to bird gatherings (except pigeon races taking place within the British Isles only, where separate conditions apply and falconry events where separate conditions apply):

1 *A nominated person must be designated as the event organizer and a person responsible for keeping the records set out in condition 2.*
2 *A record of all attendees must be kept for three months, which includes the following information:*
   a) *full name*
   b) *home address*
   c) *telephone number*
   d) *number and types of birds.*
3 *A named veterinary surgeon (or another veterinary surgeon if the named person is unavailable) must be contactable for advice and to attend in the event of any suspect disease.*
4 *Biosecurity advice must be distributed at the event.*
5 *If sales of birds are taking place at the event, a record of all sales must be kept by the event organizer for at least three months. This must include the name, address and telephone number of both the seller and the buyer, and any identifying features or individual identification of the purchased bird(s).*
6 *To notify the local State Veterinary Service Animal Health Office nearest to the event at least 14 days prior to the event taking place. Notification must include date, location, details of the event organizer, anticipated numbers and types of birds.*

BIOSECURITY BEST PRACTICE CONDITIONS TO BE APPLIED

▶ *Written detailed action plans, held by the nominated responsible person, must be available in the event of a disease incident at the event or nearby the event.*
▶ *All litter and manure within the cages, crates or baskets must be contained until disposal. Any spillages outside the cage to be cleansed and disinfected immediately.*
▶ *All litter and manure must be disposed of in a manner that does not present a risk of spread of the disease, e.g. in sealed bags for normal refuse collection in such a manner that other birds do not have direct access to it.*

▶ *All exhibitors/entrants must be instructed to cleanse and disinfect the show cages, crates or baskets before the event and be advised that they should be cleansed and disinfected on return to the home premises and before they are used to hold any other bird.*

SALES

Buyers must isolate the purchased bird(s) from any other birds (except those purchased at the same event) for at least one week. Any signs of ill health observed in the purchased bird(s) during this period must be reported to a veterinary surgeon and such birds must not be mixed with any other birds until the presence of an avian notifiable disease has been ruled out.

Contact your local State Veterinary Service and Local Authority for further advice on biosecurity measures and any other legislation that may apply.

The Great Britain Poultry register

DEFRA

The Poultry Register has been established to gather essential information about poultry held on commercial premises in Great Britain. The information in the register is to help reduce the threat or the impact of an outbreak of AI.

Why do we need a register?
If we know where your poultry are kept, and know how many birds you have, we will be able to communicate with you quickly to help protect your flocks in the event of an outbreak.

Who has to register?
There is a legal requirement* for you to register poultry if you own or are responsible for a commercial poultry premises with 50 or more poultry. This requirement applies if the premises is only usually stocked with 50 **or more** poultry for part of the year.

A commercial poultry premises means premises where poultry are kept for commercial purposes **and does not** include premises where all poultry and their eggs are kept by their owners for their own consumption or, in the case of poultry, as pets.

At present, premises with less than 50 birds are not required to register. Information on smaller flocks may be required at a later date.

*The Avian Influenza (Preventive Measures) Regulations 2005; The Avian Influenza (Preventive Measures) (Scotland) Regulations 2005; The Avian Influenza (Preventive Measures) (Wales) Regulations 2005

Which poultry must be registered?
All birds that are reared, given, sold, or kept in captivity for:

▶ *showing*
▶ *breeding*
▶ *the production of meat or eggs for consumption*
▶ *the production of other commercial products*
▶ *restocking supplies of game.*

You must register the following bird species: chickens, turkeys, ducks, geese, guinea fowl, partridges, quail, pheasants, pigeons (reared for meat only), ostriches, emus, cassowaries, rheas.

When do I have to register?
The register opened on 12 **December** 2005 and closed on 28 **February** 2006. Small flocks will be included in the register as soon as the first phase has been completed. The sooner that DEFRA knows where your poultry are, the better prepared they will be to prevent and control an AI outbreak.

Good biosecurity is vital for protecting your poultry! Information about AI and how to protect your flock is available on the DEFRA website at: www.defra.gov.uk (click on 'Avian influenza' in Quick Links).

How do I know if it is avian influenza?

SYMPTOMS

Typically the disease in poultry presents suddenly with affected birds showing oedema (swelling) of the head, cyanosis (purple/blue discoloration) of the comb and wattles, dullness, lack of appetite, respiratory distress, diarrhoea and drop in egg production. Birds may often die without any signs of disease being apparent. Death can occur so quickly that these signs may not be seen. Influenza therefore should be considered in the differential diagnosis of any significant unexplained mortality. There can be considerable variation in the clinical picture and severity of the disease. (Pictures are available on the DEFRA website at: www.defra.gov.uk/animalh/diseases/notifiable/pictures/avianinfluenza.html)

Be suspicious with:

▶ *high, rapid and unexplained mortality*
▶ *a severe drop in egg production.*

Symptoms of AI:

▶ *Large numbers of depressed, sick and dying birds*
▶ *Panting with open mouth*
▶ *Discharge from eyes and nostrils*
▶ *Dark congested comb and wattles*
▶ *Swelling of the head*
▶ *High fever*

Compare with:

▶ *Newcastle Disease: nervous signs such as twisted neck, trembling or difficulty in walking.*
▶ *Infectious Bronchitis (IB): respiratory noise, discharge from eyes and nostrils, egg production drops but not significant acute mortality.*

▶ *Mycoplasma: severe sinusitis, head swelling, sweet sickly smell, congested nostrils.*

Other viral respiratory pathogens:

▶ *Avian pneumovirus (TRT)*
▶ *Infectious laryngotracheitis (ILT)*

Bacterial respiratory pathogens:

▶ *Ornithobacter rhinotrachale (ORT)*
▶ *Haemophilus paragallinarum (fowl coryza)*
▶ *Pasturella multocida (fowl cholera)*
▶ *E. coli as secondary infection*

Other countries please contact the relevant authorities.

Taking it further

Further reading

Anderson Brown, A. and Robbins, G. E. S. (1992) *The New Incubation Book*. Fordingbridge, Hampshire, World Pheasant Association

Ashton, C. (1999) *Domestic Geese*. Crowood Press, Wiltshire

Ashton, C. and M. (2001) *The Domestic Duck*. Crowood Press, Wiltshire

Bateman, J. (1989) *Animal Traps and Trapping* (out of print, but copies are still readily available)

Beynon, P. H., Forbes, N. A., and Harcourt-Brown, N. H. (1996) *Manual of Raptors, Pigeons and Waterfowl*. BSAVA, Gloucestershire

Coles, B. H. (2007) *Essential Avian Medicine and Surgery*. Wiley-Blackwell, Oxford

Cooper, J. E. (2003) *Captive Birds in Health and Disease*. World Pheasant Association.

Hobson, J. C. (2012). *A Practical Guide to Modern Gamekeeping*. Howtobooks

Howman, K. C. R. (1980) *Pheasants, Their Breeding and Management*. World Pheasant Association, Hampshire

Robbins, G. E. S. (1981) *Quail, Their Breeding and Management.* World Pheasant Association, Hampshire

Robbins, G. E. S. (1984) *Partridges, Their Breeding and Management.* World Pheasant Association, Hampshire

Roberts, V. (1993) *Poultry at Home* (DVD). 5M Publishing, Sheffield

Roberts, V. (Ed) (2008) *British Poultry Standards.* Wiley-Blackwell, Oxford

Roberts, V. (1998) *Poultry for Anyone.* Whittet Books, Cambridgeshire

Roberts, V. (2009) *Diseases of Free-Range Poultry.* 3rd ed. Whittet Books, Cambridgeshire

Roberts, V. (2002) *Ducks, Geese and Turkeys for Anyone.* Whittet Books, Cambridgeshire

Roberts, V. (2013) *Pet Friendly Series: Chickens.* Magnet & Steel.

Roberts, V. and Scott Park, F. (2008) *Manual of Farm Pets.* BSAVA, Gloucestershire

Thear, K. (1997) *Free Range Poultry.* Whittet Books, Suffolk

Thear, K. (1990) *Keeping Quail.* Broad Leys Publishing, Essex

Van Hoesen, J. (1989) *Guinea Fowl.* USA

Wise, D. R. (1993) *Pheasant Health and Welfare.* Game and Wildlife Conservation Trust, Hampshire

WEBSITES

www.vicvet.com
The author's website for poultry husbandry, health and books.

www.animaloracle.com
Decision-tree for symptoms and timings for veterinary attention.

www.poultryclub.org
Husbandry, pure breeds, showing.

www.waterfowl.org.uk
Domestic and ornamental waterfowl.

www.vetark.co.uk
Useful products.

www.birdcareco.com
Useful products.

www.meadowsah.com
F10 disinfectant.

www.nadis.co.uk
Poultry disease information with pictures.

www.countrysmallholding.com
Poultry and *Your Chickens* magazines.

www.fancyfowl.com
Fancy Fowl magazine.

Disease Prevention: Chickens and Waterfowl

QUARANTINE (AND BIOSECURITY)

▶ *Quarantine after a show*
▶ *Quarantine new stock*
▶ *Separate quarters for 10-14 days*
▶ *Worm with Flubenvet and provide apple cider vinegar at 50ml:500ml water*
▶ *Pay attention to hand, clothes and footwear hygiene*
▶ *Avian Influenza: house stock to keep away from wild birds.*

FIVE WELFARE NEEDS

▶ *Environment – a suitable place to live*
▶ *Diet – the right food in the right amounts*
▶ *Behaviour – being able to behave normally*
▶ *Company – for animals that need to live together*
▶ *Health – protecting your pet from pain, suffering, injury and disease.*

FLOCK SIZE/COMPANY/BEHAVIOUR

▶ *More than one chicken*
▶ *A hen flock should be added all at the same time, and to fit the housing available*
▶ *Add new birds with regard to the pecking order*
▶ *Heavy breeds in pairs or trios assuming one cockerel*
▶ *Light breeds should have at least 4 hens to a cockerel*
▶ *Normal behaviour means a healthy bird*
▶ *Prey species hide symptoms of illness*
▶ *Observation is key.*

ANOREXIA

- ▶ Worm animals regularly
- ▶ Do not wait
- ▶ Hens dehydrate easily and weaken quickly
- ▶ Provide apple cider vinegar at 10ml:500ml water one week a month, or up to 50ml:500ml if unwell, for 2-3 weeks
- ▶ Use a crop tube if necessary.

VENTILATION

- ▶ Ventilation holes, slits or windows should be at the top of the house to avoid draughts
- ▶ Ventilation holes should be on two sides and never be closed
- ▶ Windows should not be glass due to risk of breakage but rather small square mesh to prevent predator entry
- ▶ Overheating is more dangerous than cold weather
- ▶ Damp bedding and high ammonia levels lead to respiratory disease.

VERMIN/PREDATOR CONTROL

- ▶ Mice are attracted by food and can get in via very small holes
- ▶ Rats are attracted by food and eggs; poison or trap them
- ▶ Grey squirrels are attracted by food and eggs and are worse in urban areas; poison or trap them
- ▶ Weasels or stoats are attracted by the birds; trap them or prevent access
- ▶ Mink are attracted by the birds; trap them or prevent access
- ▶ Foxes are attracted by the birds; prevent access
- ▶ Feral cats are attracted by the birds; prevent access
- ▶ Magpies, crows, jackdaws and rooks are attracted by food and eggs; prevent access.

Glossary

Abdomen Belly

***Ad lib* feeding** Poultry able to feed at any time (protect this from wild birds)

Airsac Air storage areas, which then act as bellows to push air through lungs

Airspace Very small when newlaid egg

Albumen White of egg

Apex roof Henhouse roof with two slopes

Automatic drinker Connected to header tank or mains with valve

Automatic Parkland feeder Feed is accessed by pecking a coloured bar

Autosexing breeds Crossbreeds that have different colours for male and female chicks

Banding Different coloured edge to feather

Bantam Very small pure breed chicken, may be also miniature of certain large fowl

Bantam ducks Small ducks, used for exhibition or slug control

Barring Horizontal band of a different colour across a feather

Batch The number of eggs set weekly in an incubator

Breed club Collection of enthusiasts for one breed

Breeder pellets Ration with extra vitamins; feed four weeks before expected eggs

Breeding pen A male and several females of one breed, selected for a good standard

British Waterfowl Association Club for domestic and wild waterfowl

Brooder Electrically heated area for young chicks, ducklings, goslings or poults

Broody Bird that sits on the nest all the time to incubate eggs

Caeca Paired blind-ended parts of intestine where some plant fermentation takes place

Call ducks Noisy, vocal, very small ducks, used for exhibition

Candle In a dark room, shine a small torch on the broad end of the egg in order to see inside

Chalazae Two strands of thick white, which help to keep the yolk the right way up (especially if fertile)

Chick Chicken up to eight weeks of age

Chick crumbs High protein small-sized feed for dayolds

Clucker A hen that is already broody

Clutch A number of eggs laid by one female until a day is missed

Cock or rooster Male over a year old

Cockerel Male chicken up to a year old

Comb Fleshy protruberance on top of head

Crate Airy poultry box usually made of plastic for hygiene

Crop Food storage area near top of breast

Crown Top of head

Domestic Waterfowl Club Club for domestic waterfowl

Down The thick underlayer of feathers, traditionally used to fill quilts

Drake Male duck

Droppings board Removable board to catch droppings placed under perches

Dual purpose Good for both eggs and meat

Duck Female duck

Duckling From dayold to six weeks

Dustbath Box with dry soil or ashes

Dustbathing A hen rolls around and flicks dry soil over her feathers to help remove parasites

Electric hen Narrow insulated box on adjustable legs with a thermostatically controlled element and a soft under material so the chicks can press up against it

Embryo Developing bird in the egg

Feed bin Vermin- and weather-proof bin such as a dustbin to keep feed in

Feeder Container for food to keep it clean and dry

Fertile eggs A male must mate with the females to fertilize eggs

Fold unit Movable self-contained house and run, may or may not have wheels

Fomites Any means of transport for pathogens, e.g. clothes, equipment

Free-range Access to grass in daylight

Free-range hybrid Commercial crossbreed, bred for outdoor conditions

Free-standing drinker Container for water to keep it clean

Germinal disc Where the yolk is fertilized, needs to stay on top, hence the chalazae

Gizzard Strong muscular organ containing small stones to grind up food

Grower chicken 8–18 weeks

Grower duck Between six weeks and adult plumage (16 weeks)

Grower pellets For growers

Hard feather Game birds, used for fighting in 1800s (cockfighting outlawed 1848)

Hatcher Separate thermostatically controlled insulated box where eggs are placed two days before hatch date, cleaned between hatches

Heat lamp Preferably infrared ceramic for safety

Heavy breeds Dual purpose, eggs and meat

Hen A laying chicken of any age or female turkey

Hock 'Elbow' of the leg

House, hut, coop, cabin Henhouse

Hybrid Commercial crossbreed, bred for indoor conditions

Imprinting Instinctive reaction in first few hours of life to follow a moving object

Incubation Keeping eggs at the correct temperature and humidity so they hatch: chickens take 21 days, ducks and guinea fowl 28, Muscovies 35, geese 28–32, quail 17

Incubator Electrically powered, thermostatically controlled, insulated box with or without automatic turning, in which eggs are incubated

Infundibulum Funnel-shaped structure where the ovum is deposited, chalazae added

Isthmus Shell membrane added here

Layer pellets For laying hens or ducks

Light breed The more active breeds

Litter Dry and friable substrate on the floor

Lungs Do not expand but have air pushed through them

Magnum Albumen (white) added here

Maintenance ration Lower protein winter feeding for breeders

Mallard Ancestor of all domestic ducks except Muscovy

Mash A commercial feed not really used for free-range as wasteful

Mixed corn Wheat and maize combined, only useful in cold weather as very heating

Mixed grit Needed for the function of the gizzard and a supply of calcium

Moult Annual replacement of feathers

Muscovy South American perching duck

Nape The back of the neck

Nestbox To lay eggs in

Oviduct Tube where the egg is constructed

Ovum (plural ova) One egg yolk in the left ovary (right ovary not functional)

Pathogens Harmful organisms

Pecking order Vital social order of poultry

Pellets A commercial ration or feed in pelleted form

Pent roof Henhouse with one slope on the roof

Perch Roosting place

Pin bones Flexible pelvic bones either side of vent

Point-of-lay (POL) 18 weeks (but in any case before laying begins and the best time to acquire chickens)

Pophole Low exit door

Post-ovulatory follicle Tissue remaining after the ovum has passed into the oviduct (ovulation)

Pot eggs Pottery or wooden eggs put in the nestbox to encourage broodiness

Poultry Club of Great Britain Responsible for Standards (www.poultryclub.org)

Preen gland Small gland just above tail producing oily substance

Preening The act of feather maintenance

Proventriculus Acid-producing stomach

Pullet A young female chicken before it lays eggs

Pure breed soft feather heavy breed Less good layer, better for meat, may be more docile

Pure breed soft feather light breed Good layer, may be nervous and flighty

Quill drinker Triangular shape, fed from header tank with nipple drinkers along base edge

Rare breed Pure breed but low in numbers in UK (does not have separate breed club)

Rump Lower back, immediately above the tail

Run or pen Fenced exercise area, usually grassed

Scapulars Waterfowl feathers in the shoulder region, shielding most of the wing when closed

Scraps Household food – this should only be given to hens if it is raw vegetable matter

Selective breeding Only breeding from those which conform to the Standard

Set To put eggs under a broody or in an incubator in order to hatch them

Sex curl The curly drake's tail (not Muscovy)

Shank Leg between hock and foot

Shavings Livestock woodshavings for litter, also to line nestbox

Sinus Area just below eye

Sitting A clutch of fertile eggs

Skids Used to move larger huts

Spiral feeder Metal spiral at base of large bucket, pellets accessed when spiral pecked

Spleen Small, purple, shiny and round

Standards Published characteristics and plumage colour for each breed

Stern Area near the vent, below the tail in waterfowl

Straw Usually wheat straw as barley straw is too soft

Sunbathing A hen lies with outstretched wing and leg towards the sun

Trachea Windpipe

True bantam Very small pure breed chicken, no large fowl counterpart

Uterus/shell gland Shell added here

Vagina Egg passes through here in one minute

Vent Anus

Ventilation Must be at roof level and above heads of birds

Vice Detrimental action

Wattles Fleshy protruberances under beak

Wheat Fed whole as a treat

Window Replace any glass with square wire mesh

Index

Page numbers in bold italic indicate illustrations.

Abacot Ranger ducks, *75–6*
abdomen, *41, 101*
ad lib feeding, *30, 33*
African geese, *122–3*
age of acquisition, *10, 84, 126, 161*
airsac, *41, **43**, 59*
airspace, *41, 45*
albumen, *41*
American Buff geese, *123–4*
ammonia, *39*
anatomy, ***10, 42–3, 100, 139, 160***
angel wing, *110–11, 148, 208*
apple cider vinegar, *246*
Araucana chickens, *44*
arks, ***16, 17***
artificial hatching, *57–9, 109–10, 147–8, 179*
artificial lighting, *16–17*
Asian hard feather breeds, *7*
Aspergillosis, *207*
automatic Parkland feeders, *30, **35***
autosexing breeds, *1, 7*
avian influenza (AI), *236–42*
avian TB, *207*
aviary system, *26, **27, 28***
Aylesbury ducks, *78–9, **86***

Bacilliary White Diarrhoea, *202*
baked eggs, *224*
banding, *154*
bantam ducks, *72, 81–2*

bantams, *2, 3, 21, 33, 46*
Bareback, *206*
Barred Wyandotte hen, ***22***
barring, *154*
bebinca (layered dessert), *225–6*
behaviour, *37–8, 46, 98–102, 136–41, 172–7*
benefits (of poultry keeping), *13–14, 87, 128–9, 163*
biosecurity guidelines, *9, 83, 125, 159–60*
Black East Indian ducks, *81–2*
Black Rock chickens, *3*
Blackhead, *172, 202*
Blue Bell chickens, *3*
Blue Swedish ducks, *79*
boiled eggs, *222*
Bourbon Red turkeys, *158*
Bovans Nera chickens, *3*
Brecon Buff geese, *121–2*
breed character, *66*
breed classification, *6–7, 74, 119*
breed clubs, *2, 73*
breeding
    chickens, *53–69*
    ducks, *106–15*
    geese, *145–51*
    turkeys, *179–80*
breeding pens, *53*
breeds, choice of
    chickens, *1–8*
    ducks, *71–83*

breeds, choice of (*Contd*)
    geese, *117–24*
    turkeys, *153–9*
British Waterfowl Association, *73*
British White turkeys, *157*
Bronze turkeys, *158–9*
brooders, *54*
broody coups, **56**
broody ducks, *73, 105, 108–9*
broody geese, *143–4, 145*
broody hens, *2, 5, 51–2, 54, 55–7*
broody turkeys, *177–8*
brown eggs, *44*
Buff Back geese, *122*
Buff Orpington ducks, *76*
Buff turkeys, *157*
bullying, *37–8*
Bumblefoot, *205*
buying poultry, *9–11, 83–4, 125–6, 159–61*

caeca, *41,* **42**
Calder Ranger chickens, *3*
Call ducks, *72, 73, 82–3, 98*
candling, *41, 45, 59–60, 110*
canker, *208*
cannibalism, *16, 37, 172*
caruncles, *73*
caudal thoracic airsac, **43**
Cayuga ducks, *79*
cervical airsac, **43**
chalazae, *41*
chick crumbs, *30, 56, 63, 111*
chick discomfort, *210*
chick feeders, **58**
chicken breeds, *1–8*
chickens, meat production, *189–90*
chicks, *2, 189, 211*

Chinese geese, *119–20*
chocolate raspberry roulade, *226–7*
classification of breeds, *6–7, 74, 119, 155*
clavicular airsac, **43**
cleaning
    eggs, *44*
    housing, *21*
    incubators, *60*
    poultry, *194–7*
cluckers, *54*
clutches, *2, 73*
coccidiosis, *63, 202, 212–15*
cockerels, *47–8, 189–90*
colour vision, *31, 37, 57, 98, 99, 137, 172*
Columbian Blacktail chickens, *3*
combs, *41, 66*
commercial hybrids, *8*
conformation, *65–6*
cooking, *221–8*
costs, *13, 86–7, 128, 162–3*
coverts, *73*
cranial thoracic, **43**
crates, **12**
creosote, *19*
Crested ducks, *76*
crop, *41,* **42**
Cropbound, *207*
crumpet breakfast, *224*
culling, *68–9*
culmen, *118*
Cuprinol, *19*

dayolds, *54, 55, 57, 108, 109, 147, 179*
deaths, *51, 105, 143, 177*
decoy ducks, *82–3*

deep litter system, *19–20, 21, 25–6,* **28***, 169*
DEFRA, *52–3, 74, 237*
diarrhoea, *210*
digestive system, *30–1,* **31***, 96*
diseases, *38–9, 202–9, 211–20*
disinfectants, *21, 57, 60, 92*
disinfecting procedures, *9, 83*
Domestic Waterfowl Club, *73*
down, *101*
drakes, *73, 99*
drake's tail, *73*
drinkers, *13, 30, 32, 33,* **34***,* **36***,* **58***, 61, 89,* **97**
droppings boards, *14, 15, 18, 19, 165*
duck breeds, *71–83*
duck eggs, *74*
duck houses, **91***,* **94**
Duck viral enteritis, *209*
ducklings, *73, 106, 110–13*
dustbathing, *41, 47*

ear infections, *208*
eclipse, *73*
egg bound, *208*
egg eating, *50, 104, 176–7*
Egg Marketing Inspectorate, *52–3*
egg production, *3, 13, 101–2, 128, 163*
egg sanitizers, *60*
eggs
    cooking, *221–8*
    duck, *74, 101*
    freshness, *45*
    goose, *117*
    problems, *211*
    selling, *52–3, 105–6, 144, 178*

electric hen brooders, *54,* **58***, 62*
Embden geese, *124*
embryos, *54*
Enteritis, *205*
exhibiting, *69, 233–5*
exhibition strains, *5, 67*
eyestripe, *73*

feather pecking, *16, 47, 51, 61, 177, 210*
feather shapes, **64**
feed bins, *30*
feeders, *13, 30, 33,* **34–5***,* **97**
feeding
    chickens, *30–6*
    chicks, *61–2*
    ducks, *89–90, 96–7*
    geese, *130, 135–6*
    technical terms, *30*
    turkeys, *170–1*
fencing, fox-proof, **230**
feral cats, *229*
fertile eggs, *44–5, 54, 55, 67–8, 107, 114*
fertility, *67–8, 114–15, 151*
flock mentality, *99, 137, 174*
floors, *20, 92, 132, 166–7*
Flubenvet, *48–9, 102–3, 141–2, 246*
fold units, *14, 22, 93*
food, *33*
foxes, *15, 18, 229,* **230**
free-range eggs, *52–3*
free-range hybrids, *2, 8*
free-range poultry, *14, 21–2, 92, 167–8*

gander, *118*
geese breeds, *117–24*

geese, meat production, *190*
general licence conditions, *237–9*
gizzard, *30, 33, 41*, **42**
gizzard worm, *103, 142*
goose house, **91**
goslings, *118, 145, 148–9*
Great Britain Poultry Register, *239–41*
greenery, *13, 33, 34*
grit, *13, 33, 97, 136*
growers, *2, 73, 106, 110–11, 211–12*
guinea fowl, *182–4*, **183**, *190*
gullet, *118*
Gumboro (IBD), *207*
gutting, *194–7*, **195, 196**

hand washing, *9, 83*
handling poultry, *11–12, 85, 127, 161–2*
hanging, *192–4*
hard feather breeds, *2, 7*
hat feeder, **34**
hatchers, *54*
hatching
    chickens, *54–60*
    ducks, *107–10*
    geese, *146–8*
    turkeys, *179*
hay, *19, 92*
head points, *66*
health, signs of, *9, 65, 83–4, 125, 160*
heart, **42**
heart failure, *205*
heat lamps, *54, 61–2, 111*
heavy breeds
    chickens, *3, 33*
    ducks, *73, 78–81*

geese, *122–4*
    turkeys, *158–9*
holidays, *53*
housing
    chickens, *13–28*
    ducks, *87–94*, **91**
    geese, *129–33*
    technical terms, *14, 87–8*
    turkeys, *163–9*
humane killer, **191**
huts
    see housing
Hy-line chickens, *3*
hybrids, *2, 3, 8, 74*

import/export regulations, *9*
imprinting, *107, 110, 148*
income, *13–14, 87, 128–9, 163*
incubators, *54, 57–9, 60*
Indian breakfast omelette, *223–4*
Indian Runner ducks, *76–7*
Infectious Bronchitis (IB), *207, 219, 241*
infundibulum, *41*, **43**
injuries, *206*
Isabrown chickens, *3*
isthmus, *41*, **43**

Japanese quail, *184–6*, **185**

kedgeree, *225*
Khaki Campbell ducks, *77*

larders, fly-proof, *193*
Lavender turkeys, *157*
layered dessert (bebinca), *225–6*
laying, *3, 44, 46, 101–2, 112, 140*

laying capabilities, *4–5, 75, 118, 155*
leg colour, *46*
legislation, avian influenza (AI), *237–40*
lice, *49, 175–6, 203*
licence conditions, *237–9*
life expectancy, *2, 212*
light, *16–17, 46*
light breeds
    chickens, *3, 33*
    ducks, *73, 75–8*
    geese, *119–21*
    turkeys, *156*
Light Sussex hen, **6**
litter, *14, 19–20, 92, 166*
louse powders, *38*

Magpie ducks, *77*
magpies, *210, 229, 230*
maize, *33–4*
mandibles, *118*
manure, *14*
Maran chickens, *63*
Marek's disease, *206, 215–16*
mash, *30, 33*
materials, houses, *19, 91, 131–2, 165*
mating, *174–5*
mating up, *66–7, 114, 150*
meal, *33*
meat production, *188–99*
meat spot, *45*
medicines, *38–9*
medium breeds, geese, *121–2*
meringues, *222*
mice, *18, 39, 173, 229*
mites, *49, 103, 142, 176, 203*

mixed corn, *30*
mixed grit, *30, 97*
moulting, *41, 47, 102, 140–1, 174*
Muscovy ducks, *73, 74, 79–80, 85, 90, 104, 107*
mustard chilli eggs, *228*
mycoplasma, *38, 98, 216–19, 242*
Mycoplasmosis, *203, 204*

Narragansett turkeys, *159*
natural hatching, *55–7, 108–9, 146–7*
Nebraskan Spotted turkeys, *157*
neck dislocation, *68, 69, 191*
neck-feather partings, *118*
neck ring, *73*
nestboxes, *14, 15, 17, 19, 90, 131, 164*
Newcastle Disease, *241*
nitrate poisoning, *205*
Norfolk Black turkeys, *157–8*
northern fowl mites, *49, 103–4, 142–3, 176*

Old English Game breed, *38*
omelettes, *223–4*
over mating, *104, 143*
overgrowth, *206*

Paracox, *63*
parasites
    chickens, *40, 47, 48–9*
    duck, *102–4*
    geese, *141–3*
    turkeys, *175–6*
parasitic worms, *202*
parent stock selection, *65–6*
Parkland automatic feeder, **35**

Partridge Wyandotte hen, *40*
paunch, *118*
pecking, feather, *47, 51, 61, 177*
pecking order, *2, 5, 37–8, 99, 138*
Pekin ducks, **72**, *80*
pellets, *30, 33, 54, 63*
pens, *14, 21–2*
perches, *14, 15, 16, 17, 18, 62, 90, 165*
Pied (Cröllwitzer) turkeys, *158*
pigmentation, of eggs, *44*
Pilgrim geese, *120*
plucking, *192–4*
poached eggs, *222–3*
point-of-lay (POL), *2, 10*
poison, *230–1*
poisonous plants, *22, 92–3*
Pomeranian geese, *122*
ponds, *88*
popholes, *14, 18, 90–1, 131, 165*
poult, *154*
Poultry Club of Great Britain, *2, 67*
preen gland, *41, 102*
preening, *47*
processing, *194–7*
pseudomonas, *204*
pullets, *2, 10*
pure breeds, *2, 3–8, 65–9*

quail, *184–6*, **185**, *190*
quarantine, *246*
quill drinkers, *30,* **36**

rain, *38*
rare breeds, *2, 7*
rats, *18, 20, 39, 173, 210, 229*
rearing poultry, *60–3, 110–13, 148–9, 179–80*

recipes, *222–8*
red mites, *15, 19, 49, 103, 142, 176*
regulations, *1, 52–3, 74, 237–9*
reproductive system, **43**, *44–5*
respiratory noises, *210*
respiratory system, *42,* **43***, 98*
Rhode Island Red hen, **32**
ringing scheme, *67*
rodents *see* mice; rats
Roman geese, *120*
roofing felt, *15, 19*
roofs, *14, 15, 19*
Rouen ducks, *80–1*
routines
    artificial rearing, *62–3*
    chickens, *29, 37*
    ducks, *95*
    geese, *134*
    turkeys, *169–70*
runs, *14,* **20***, 21–2*

sales, *239*
Saxony ducks, *81*
scaly leg mites, *49, 176, 204*
scrambled eggs, *223*
scraps, *30, 135–6*
Sebastopol geese, *120–1*
security, *18*
selection
    parent stock, *65–6*
    see also breeds, choice of
selective breeding, *54*
selling
    eggs, *13, 52–3, 105–6, 144, 178*
    surplus stock, *69*
semen storage, *44, 173*
sexing, *63–4, 113, 149–50*

shavings, *13, 14, 19–20, 92*
showing, *69, 233–5*
Silkies, *64*
Silver Appleyard ducks, *81*
Silver Appleyard miniature
 ducks, *82*
Silver Bantam ducks, *82*
Silver Dutch Cockerel, *50*
Silver Grey Dorking, *64*
skinning, *194*
Slate turkeys, *158*
slaughter, *191–2*
soft feather breeds, *2, 6*
space
 chickens, *15–16, 52–3*
 ducks, *88*
 geese, *129–30*
 turkeys, *164*
Speckledy chickens, *3*
spiral feeders, *30, **35***
Standards, *2, 3, 65–6, 73*
Steinbacher geese, *121, **123***
storage bins, *13, 30, 34*
straw, *13, 14, 19, 92*
stress, *98, 206*
string, *39, 102, 174*
stunning, *191*
sudden death, *211*
sunbathing, *42, 47*
swimming, *109*
syrinx, *42, 43*

tarragon baked eggs, *224*
temperature, *40*
Toulouse geese, *124*
transport recommendations, *232*
traps, *229*
trios, *2, 73*

troughs, **97**
Trout Indian Runners, **78**
true bantams, *2, 7*
trussing, *197, **197–8***
tumours, *206*
turkey breeds, *155–9*
Turkey Club, *154, 156*
turkeys, meat production, *190*

vaccination, *38*
vegetable matter, *33*
 see also greenery
vent-sexing, *63, 113, 149–50*
ventilation, *14, 15, 16–17, 42, 90,*
 131, 164
vermin, *15, 18, 165, 229–31*
vices
 chickens, *42, 50–1*
 ducks, *104*
 geese, *143*
 turkeys, *176–7*
Virkon, *21, 57, 60, 92*

Warren chickens, *3*
washing eggs, *44*
water, *88–9, 130*
waterfowl lice, *103, 142*
waterfowl, meat production, *190*
watering, *30–6, 96–7, 135–6,*
 170–1
weight loss, *210*
welfare, *36–7, 98–105, 136–43,*
 172–7
Welsh Harlequin ducks, *77–8*
Welsummers, *63*
Wet feather, *209*
wheat, *30, 33–4*
wheat straw, *92*

White Campbell ducks, *104*
White Star chickens, *3*
wild bird access, *21, 132, 167*
windows, *14, 16–17, 164*
wing clipping, *47, **48**, 141*
wing joint problems, *110–11, 148*
withdrawal ration, *188*

wood treatments, *19*
woodshavings *see* shavings
worming, *38, 48–9, 102–3, 141–2, 173*
wounds, *205*

yolk colour, *45*

Notes

Notes

Notes

Notes